V. Loia, M. Nikravesh, L. A. Zadeh (Eds.)

Fuzzy Logic and the Internet

Springer

Berlin
Heidelberg
New York
Hong Kong
London
Milano
Paris
Tokyo

Studies in Fuzziness and Soft Computing, Volume 137

Editor-in-chief
Prof. Janusz Kacprzyk
Systems Research Institute
Polish Academy of Sciences
ul. Newelska 6
01-447 Warsaw
Poland
E-mail: kacprzyk@ibspan.waw.pl

Vincenzo Loia
Masoud Nikravesh
Lotfi A. Zadeh (Eds.)

Fuzzy Logic
and the Internet

Springer

Prof. Vincenzo Loia
Universita di Salerno
Dipto. Matematica e Informatica
Via S. Allende
84081 Baronissi
Italy
E-mail: loia@unisa.it

Prof. Masoud Nikravesh
E-mail: nikravesh@cs.berkeley.edu

Prof. Dr. Lotfi A. Zadeh
E-mail: zadeh@cs.berkeley.edu

University of California
Dept. Electrical Engineering and Computer
Science - EECS
94720 Berkeley, CA
USA

ISSN 1434-9922
ISBN 3-540-20180-7 Springer-Verlag Berlin Heidelberg New York

Library of Congress Cataloging-in-Publication-Data

Fuzzy logic and the Internet / Vincenzo Loia, Masoud Nikravesh, Lotfi A. Zadeh (eds.).
p. cm.
ISBN 3-540-20180-7 (alk. paper)
1. Fuzzy logic. 2. Internet research. 3. Internet searching. I. Loia, Vincezo, 1961- II.
Nikravesh, Masoud, 1959- III. Zadeh, Lotfi Asker.
QA76.87.F895 2004
004.67'8--dc22

Springer-Verlag is a part of Springer Science+Business Media
springeronline.com

© Springer-Verlag Berlin Heidelberg 2004
Printed in Germany

Typesetting: Camera-ready by author
Cover design: E. Kirchner, Springer-Verlag, Heidelberg
Printed on acid free paper   62/3020/M - 5 4 3 2 1 0

Preface

With the daily addition of million documents and new users, there is no doubt that the World Wide Web (WWW or Web shortly) is still expanding its global information infrastructure. Thanks to low-cost wireless technology, the Web is no more limited to homes or offices, but it is simply everywhere. The Web is so large and growing so rapidly that the 40 million page "WebBase" repository of Inktomi corresponds to only about 4% of the estimated size of the publicly indexable Web as of January 2000 and there is every reason to believe these numbers will all swell significantly in the next few years.

This unrestrainable explosion is not bereft of troubles and drawbacks, especially for inexpert users. Probably the most critical problem is the effectiveness of Web search engines: though the Web is rich in providing numerous services, the primary use of the Internet falls in emails and information retrieval activities. Focusing in this latter, any user has felt the frustrating experience to see as result of a search query overwhelming numbers of pages that satisfy the query but that are irrelevant to the user.

Due to nature of the Web itself, there is a strong need of new research approaches, in term of theories and systems. Among these new research trends, an important role is played by those methodologies that enable to process imprecise information, and to perform approximate reasoning capability. The ability of Fuzzy Technology to exploit the tolerance for imprecision to achieve tractability, robustness, and low solution cost, has played a fundamental and successful role in any area of Information Technology, reporting a growing interest especially in the area of computational intelligence.

Nowadays Web-based systems handle user interaction in matching user's queries that are too weak to cope with the user's expressiveness. First attempts in extending searching towards deduction capability are essentially based on two-valued logic and standard probability theory. The complexity of the problem coupled with some features of the space domain (unstructured data, immature standards) demand a strong deviation from this trend. Fuzzy Logic, and more in general Soft Computing, can be a right choice to face complex Web problems, as reported by the contributions of this volume.

This book contains 14 chapters.

First chapter written by Nikravesh, Takagi, Tajima, Loia, and Azvine is an introduction to the book. The main objective of this chapter to provide a better understanding of the issues related to the Internet (Fuzzy Logic and the Internet), and provide new tools and ideas Toward the Enhancing the Power of the Internet. The main purpose of this chapter to draw the attention of the fuzzy logic community as well as the Internet community to the fundamental importance of specific Internet-related problems. This issue is critically significant about problems that center on search and deduction in large, unstructured knowledge bases. The authors summarize the challenges, the road ahead and directions for the future by recognizing the challenging problems and the new direction toward the next generation of the search engines and Internet.

Chapter 2 written by Beg and Ahmad is on the problem of rank aggregation on the web. In this chapter the authors propose new ranking solutions, namely MFO, MBV, Improved Shimura and Entropy-based ranking. MFO works by comparing the values of the membership functions of the document positions. MBV proceeds by carrying out an ascending sort on the ratio of mean and variance of the document positions. The Shimura technique is improved by replacing the min function with the OWA operation. The entropy-based technique goes about by adopting the entropy minimization principle for the purpose of rank aggregation.

Chapter 3 written by Cordón, Moya and Zarco explains how it is possible to automatically derive extended Boolean queries for fuzzy information retrieval systems from a set of relevant documents provided by a user. The chapter features an advanced evolutionary algorithm, GA-P, specially designed to tackle with multi-objective problems by means of a Pareto-based multi objective technique. The approach is experimented on the usual Cranfield collection and compared to other well-known methods.

Chapter 4 written by Damiani, Lavarini, Oliboni, and Tanca proposes a flexible querying technique, XML compliant, able to locate and extract information. In structure and tag vocabulary. This approach relies on representing XML documents as graphs, whose edges are weighted at different levels of granularity. A smart weighting technique process the features of the edges, generating a separate weight according to each characteristic, and then aggregating these values in a single arc-weight. An important optimization is carried out by a threshold-based pruning that deletes unimportant edges, in order to retain only the most useful information for an efficient Web searching strategy.

Chapter 5 written by Herrera , Herrera-Viedma, Martínez and Porcel describes a distributed intelligent model for gathering information on the Internet, where the agents and users may communicate among them using a multi-granular linguistic technique based on a linguistic 2-tuple computational. Different advantages derive from this technique: the retrieval process gains in flexibility, the agent-oriented interaction can benefit from a deeper expressivity, and the availability of a words-based computation improves precision without loss of information.

Chapter 6 written by Hong, Lin and Wang presents a fuzzy web-mining algorithm for processing web-server logs in order to discover fuzzy browsing patterns among them. The chapter describes how this approach can derive a more complete set of browsing patterns than other previous solutions, detailing some experimental results for showing the time-completeness trade-off effects.

Chapter 7 written by Liu, Wan and Wang describes a fuzzy inference system for audio classification and retrieval, a crucial problem for any multimedia Web search engine. The chapter illustrates the benefits of the fuzzy classifier that is characterized by a very quick classification, flexibility and efficiency in adding new classes to audio samples in the database.

Chapter 8 written by Loia is on Web searching catalogues. In many cases, these catalogues are maintained manually with enormous costs and difficulty due to the incessant growing of the Web. The chapter presents an evolutionary approach useful to construct automatically the catalogue as well as to perform the classification of a Web document. This functionality is achieved by a genetic-based fuzzy clustering applied on the context of the document, as opposite to content-based clustering that works on the complete document information.

Chapter 9 written by Martín-Bautista, Sánchez, Serrano and Vila addresses the problem of query specification by describing an application of data mining techniques in a text framework. The chapter proposes a text transaction technology based on fuzzy transactions, considering that each transaction correspond to a document representation. The set of transactions represents a document collection from which the fuzzy association rules are extracted. The extracted can be automatically added to the original query in order to optimize the search.

Chapter 10 written by Nikravesh and Azvine introduces fuzzy query and fuzzy aggregation as an alternative for ranking and predicting the risk for credit scoring and university admissions. The chapter presents the BISC Decision Support System characterized by smart Internet-based services designed to use intelligently the vast amounts of important data in complex organizations and to share internal data with external entities by respecting the constraints of security and efficiency.

Chapter 11 written by Pal, Talwar and Mitra provides an overview on different characteristics of web data, the basic components of web mining and its different types, and their current states of the art. The chapter underlines the limitations existing in web mining methods and evidenciate how the soft computing approach can be a valid ally to achieve Web intelligence.

Chapter 12 written by Pasi and Yager presents a technique suitable to improve the quality of the information available to customers in making Web purchase decisions. The Product Category Summarization (PCS) method is presented, and the chapter illustrates who PCS is able to help the consumers understanding a product line in a way that can help them in their purchasing decisions. PCS, after providing a clustering of a product line into a finite number of categories, automatically constructs some user friendly descriptions of the relevant features shared by the majority of the products associated with each category.

Chapter 13 written by Pham faces logo technology, widely used nowadays to meet an increasing demand for the automatic processing of documents and images. The chapter outlines the concept of geostatistics that serves as a tool for extracting spatial features of logo images. Different logo classifiers experiences are discussed, ranging from a model based on neural networks, pseudo hidden Markov models, and fuzzy sets, up to an algorithm built-on the concept of the mountain clustering.

Chapter 14 written by Wang and Zhang presents a Fuzzy web information classification agent based on Fuzzy Web Intelligence. The agent can act upon user's instructions and refresh the stock data in a real time manner by accessing the database on the Internet. Using fuzzy reasoning, the agent can create a list of top stocks based on the output values calculated from input stock information. The chapter shows how the results of the data processing are precise and reliable.

We thank the authors for their outstanding contribution in this book. Thanks are due to Professor J. Kacprzyk for his kind support that encouraged us in preparing this volume. We are also very grateful to the editorial team of the Springer-Verlag Company for the continuous fruitful assistance.

Vincenzo Loia
Masoud Nikravesh
Lotfi Zadeh

Table of Contents

Fuzzy Logic and the Internet: Web Intelligence

Masoud Nikravesh[1], Tomohiro Takagi[2], Masanori Tajima[2], Vincenzo Loia[3], and Ben Azvine[4]

[1]BISC Program, EECS Department-CS Division
University of California, Berkeley, CA 94720
Nikravesh@cs.berkeley.edu
[2] Dept. of Computer Science, Meiji University.
[3]Dipartimento di Matematica e Informatica
Università di Salerno
84081 Baronissi (Salerno) ITALY
[4]BTExact Technologies
Orion Building pp1/12, Adastral Park,
Martlesham, Ipswich IP5 3RE, UK

Abstract: Retrieving relevant information is a crucial component of cased-based reasoning systems for Internet applications such as search engines. The task is to use user-defined queries to retrieve useful information according to certain measures. Even though techniques exist for locating exact matches, finding relevant partial matches might be a problem. It may not be also easy to specify query requests precisely and completely - resulting in a situation known as a fuzzy-querying. It is usually not a problem for small domains, but for large repositories such as World Wide Web, a request specification becomes a bottleneck. Thus, a flexible retrieval algorithm is required, allowing for imprecise or fuzzy query specification or search.

1 Introduction

Humans have a remarkable capability (perception) to perform a wide variety of physical and mental tasks without any measurements or computations. Familiar examples of such tasks are: playing golf, assessing wine, recognizing distorted speech, and summarizing a story. The question is whether a special type information retrieval processing strategy can be designed that build in perception.

World Wide Web search engines have become the most heavily-used online services, with millions of searches performed each day. Their popularity is due, in part, to their ease of use. The central tasks for the most of the search engines can be summarize as 1) query or user information request- do what I mean and not what I say!, 2) model for the Internet, Web representation-web page collection, documents, text, images, music, etc, and 3) ranking or matching function-degree of relevance, recall, precision, similarity, etc. *Table 1* also compares the issues related to the conventional Database with Internet. Already explosive amount of users on the Internet is estimated over 200 million (*Table 2)*. While the number of pages available on the Internet almost double every year, the main issue will be the size of the internet when we include multimedia information as part of the Web and also when the databases connected to the pages to be considered as part of an integrated Internet and Intranet structure. Databases are now considered as backbone of most of the E-commerce and B2B and business and sharing information through Net between different databases (Internet-Based Distributed Database) both by user or clients are one of the main interest and trend in the future. In addition, the estimated user of wireless devices is estimated 1 billion within 2003 and 95 % of all wireless devices will be Internet enabled within 2005. *Table 3.* shows the evolution of the Internet, World Wide Web, and Search Engines.

Table 1. Database Vs. Internet

Database	**Internet**
Distributed	Distributed
Controlled	Autonomous
Query (QL)	Browse (Search)
Precise	Fuzzy/Imprecise
Structure	Unstructured

Table 2. Internet and rate of changes

Jan 1998: 30 Millions web hosts
Jan 1999: 44 Millions web hosts
Jan 2000: 70 Millions web hosts
Feb 2000: +72 Millions web hosts

Dec 1997: 320 Millions
Feb 1999: 800 Millions
March 2000: +1,720 Millions

The number of pages available on the Internet almost doubles every year

Courtois and Berry (Martin P. Courtois and Michael W. Berry, ONLINE, May 1999-Copyright © Online Inc.) published a very interesting paper "Results Ranking in Web Search Engines". In their work for each search, the following topics were selected: credit card fraud, quantity theory of money, liberation tigers, evolutionary psychology, French and Indian war, classical Greek philosophy, Beowulf criticism, abstract expressionism, tilt up concrete, latent semantic indexing, fm synthesis, pyloric stenosis, and the first 20 and 100 items were downloaded using the search engine. Three criteria 1) All Terms, 2) Proximity, and 3) Location were used as a major for testing the relevancy ranking. *Table 4* shows the concept of relevancy and its relationship with precision and recall (*Table 5* and *Figure 1*). The effectiveness of the classification is defined based on the precision and recall (*Tables 4-5* and *Figure 1*).

Table 4. Similarity/Precision and Recall

	Relevant	Non-Relevant	
Retrieved	$A \cap B$	$\bar{A} \cap B$	B
Not Retrieved	$A \cap \bar{B}$	$\bar{A} \cap \bar{B}$	$\bar{B}$
	A	$\bar{A}$	N

N: Number of documents

Table 5. Similarity/Measures of Association

There are five commonly used measures of association in IR :

Simple matching Coefficiet : $|X \cap Y|$

Dice' s Coefficiet : $2\dfrac{|X \cap Y|}{|X| + |Y|}$

Jaccard' s Coefficiet : $\dfrac{|X \cap Y|}{|X \cup Y|}$

Cosine Coefficiet : $\dfrac{|X \cap Y|}{|X|^{1/2} |Y|^{1/2}}$

Overlap Coefficiet : $\dfrac{|X \cap Y|}{min(|X|,|Y|)}$

Disimilarity Coefficeint : $\dfrac{|X \cap Y|}{|X| + |Y|} = 1 - Dice' s Coefficient$

$|X \cup Y| = |X| + |Y| - |X \cap Y|$

Effectiveness is a measure of the system ability to satisfy the user in terms of the relevance of documents retrieved. In probability theory, precision is defined as conditional probability, as the probability that if a random document is classified under selected terms or category, this decision is correct. Precision is defined as portion of the retrieved documents that are relevant with respect to all retrieved documents; number of the relevant documents retrieved divided by all documents retrieved. Recall is defined as the conditional probability and as the probability if a random document should be classified under selected terms or category, this decision is taken. Recall is defined as portion of the relevant retrieved documents that are relevant with respect to all relevant documents exists; number of the relevant documents retrieved divided by all relevant documents. The performance of each request is usually given by precision-recall curve (*Figure 1*). The overall performance of a system is based on a series of query request. Therefore, the performance of a system is represented by a precision-recall curve, which is an average of the entire precision-recall curve for that set of query request.

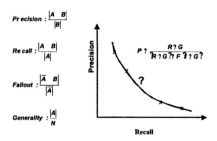

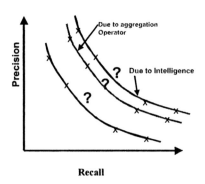

Figure 1. **1.a.)** relationship between Precision and Recall, **1.b.)** inverse relationship between Precision and Recall.

4

To improve the performance of a system one can use different mathematical model for aggregation operator for (A∩B) such as fuzzy logic. This will sift the curve to a higher value as is shown in *Figure 1.b*. However, this may be a matter of scale change and may not change the actual performance of the system. We call this improvement, virtual improvement. However, one can shit the curve to the next level, by using a more intelligent model that for example have deductive capability or may resolve the ambiguity (*Figure 1.b*).

Many search engines support Boolean operators, field searching, and other advanced techniques such as fuzzy logic in variety of definition and in a very primitive ways. While searches may retrieve thousands of hits, finding relevant partial matches and query relevant information with deductive capabilities might be a problem. *Figure 2*. shows a schematic diagram of model presented by Lotfi A. Zadeh (2002) for the flow of information and decision. What is also important to mention for search engines is query-relevant information rather than generic information. Therefore, the query needs to be refined to capture the user's perception. However, to design such a system is not trivial, however, Q/A systems information can be used as a first step to build a knowledge based to capture some of the common user's perceptions. Given the concept of the perception, new machineries and tools need to be developed. Therefore, we envision that non-classical techniques such as fuzzy logic based-clustering methodology based on perception, fuzzy similarity, fuzzy aggregation, and FLSI for automatic information retrieval and search with partial matches are required.

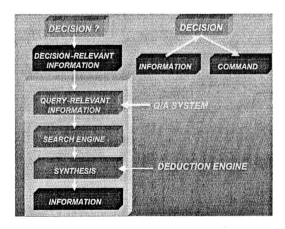

Figure 2. Perception-Based Decision Analysis (PDA) (Zadeh 2001)

2 Intelligent Search Engines

Design of any new intelligent search engine should be at least based on two main motivations:

- The web environment is, for the most part, unstructured and imprecise. To deal with information in the web environment what is needed is a logic that supports modes of reasoning which are approximate rather than exact. While searches may retrieve thousands of hits, finding decision-relevant and query-relevant information in an imprecise environment is a challenging problem, which has to be addressed.
- Another, and less obvious, is deduction in an unstructured and imprecise environment given the huge stream of complex information.

Tim Bernres-Lee (1999) in his transcript refers to the fuzzy concept and the human intuition with respect to the Web (Transcript of Tim Berners-Lee's talk to the LCS 35th Anniversary celebrations, Cambridge Massachusetts, 1999/April/14):

Lotfi A. Zadeh (2001a) consider fuzzy logic is a necessity to add deductive capability to a search engine: "Unlike classical logic, fuzzy logic is concerned, in the main, with modes of reasoning which are approximate rather than exact. In Internet, almost everything, especially in the realm of search, is approximate in nature. Putting these two facts together, an intriguing thought merges; in time, fuzzy logic may replace classical logic as what may be called the brainware of the Internet.
...
In my view, among the many ways in which fuzzy logic may be employed, there are two that stand out in importance. The first is search. Another, and less obvious, is deduction in an unstructured and imprecise environment. Existing search engines have zero deductive capability. ... To add a deductive capability to a search engine, the use of fuzzy logic is not an option - it is a necessity."

With respect to the deduction and its complexity, Lotfi's viewpoint [2001a and 2002] is summarized as follows:
"Existing search engines have many remarkable capabilities. But what is not among them, is the deduction capability -- the capability to answer a query by drawing on information which resides in various parts of the knowledge base or is augmented by the user. Limited progress is achievable through application of methods based on bivalent logic and standard probability theory. But to move beyond the reach of standard methods it is necessary to change direction. In the approach, which is outlined, a concept which plays a pivotal role is that of a prototype -- a concept which has a position of centrality in human reasoning, recognition, search and decision processes. ... The concept of a prototype is in-

trinsically fuzzy. For this reason, the prototype-centered approach to deduction is based on fuzzy logic and perception-based theory of probabilistic reasoning, rather than on bivalent logic and standard probability theory. What should be underscored, is that the problem of adding deduction capability to search engines is many-faceted and complex. It would be unrealistic to expect rapid progress toward its solution."

During 80, most of the advances of the automatic document categorization and IR were based on knowledge engineering. The models were built manually using expert systems capable of taking decision. Such expert system has been typically built based on a set of manually defined rules. However, the bottleneck for such manual expert systems was the knowledge acquisition very similar to expert system. Mainly, rules needed to be defined manually by expert and were static. Therefore, once the database has been changed or updated the model must intervene again or work has to be repeated anew if the system to be ported to a completely different domain. By explosion of the Internet, these bottlenecks are more obvious today. During 90, new direction has been merged based on machine learning approach. The advantage of this new approach is evident compared to the previous approach during 80. In machine learning approach, most of the engineering efforts goes towards the construction of the system and mostly is independent of the domain. Therefore, it is much easier to port the system into a new domain. Once the system or model is ported into a new domain, all that is needed is the inductive, and updating of the system from a different set of new dataset, with no required intervention of the domain expert or the knowledge engineer. In term of the effectiveness, IR techniques based on machine learning techniques achieved impressive level of the performance and for example made it possible automatic document classification, categorization, and filtering and making these processes viable alternative to manual and expert system models.

Doug B. Lenat both the founder of the CYC" project and president of Cycorp [http://www.cyc.com] puts the concept of deduction into perspective and he expresses that both commonsense knowledge and reasoning are key for better information extraction (2001).

Lotfi A. Zadeh (2002) express qualitative approach towards adding deduction capability to the search engine based on the concept and framework of proto-forms:
"At a specified level of abstraction, propositions are p-equivalent if they have identical protoforms." "The importance of the concepts of protoform and p-equivalence derives in large measure from the fact that they serve as a basis for knowledge compression."
"A knowledge base is assumed to consist of a factual database, FDB, and a deduction database, DDB. Most of the knowledge in both FDB and DDB is perception-based. Such knowledge cannot be dealt with through the use of bivalent logic and standard probability theory. The deduction database is assumed to consist of a logical database and a computational database, with the rules of deduction having

the structure of protoforms. An example of a computational rule is "if Q_1 A's are B's and Q_1 (A and B)'s are C's," then "$Q_1 Q_2$ A's are(B and C)'s, where Q_1 and Q_2 are fuzzy quantifiers and A, B and C are labels of fuzzy sets. The number of rules in the computational database is assumed to be very large in order to allow a chaining of rules that may be query-relevant."

Computational theory of perception (CTP) (Zadeh 1999 and 2001b, Nikravesh et al. 2001, Nikravesh,2001a and 2001b) is one of the many ways that may help to address some of the issues presented by both Berners Lee and Lotfi A. Zadeh earlier, a theory which comprises a conceptual framework and a methodology for computing and reasoning with perceptions. The base for CTP is the methodology of computing with words (CW) (Zadeh 1999). In CW, the objects of computation are words and propositions drawn from a natural language.

3 Fuzzy Logic and the Internet

During the recent years, applications of fuzzy logic and the Internet from Web data mining to intelligent search engine and agents for Internet applications have greatly increased (Nikravesh and Azvine 2001). Martin (2001) concluded that semantic web includes many aspects, which require fuzzy knowledge representation and reasoning. This includes the fuzzification and matching of concepts. In addition, it is concluded that fuzzy logic can be used in making useful, human-understandable, deduction from semi-structured information available in the web. It is also presented issues related to knowledge representation focusing on the process of fuzzy matching within graph structure. This includes knowledge representation based on conceptual graphs and Fril++. Baldwin and Morton (1985) studied the use of fuzzy logic in conceptual graph framework. Ho (1994) also used fuzzy conceptual graph to be implemented in the machine-learning framework. Baldwin (2001) presented the basic concept of fuzzy Bayesian Nets for user modeling, message filtering and data mining. For message filtering the protoype model representation has been used. Given a context, prototypes represent different types of people and can be modeled using fuzzy rules, fuzzy decision tree, fuzzy Bayesian Net or a fuzzy conceptual graph. In their study, fuzzy set has been used for better generalization. It has been also concluded that the new approach has many applications. For example, it can be used for personalization of web pages, intelligent filtering of the Emails, providing TV programs, books or movie and video of interest. Cao (2001) presented the fuzzy conceptual graphs for the semantic web. It is concluded that the use of conceptual graph and fuzzy logic is complementary for the semantic web. While conceptual graph provide a structure for natural language sentence, fuzzy logic provide a methodology for computing with words. It has been concluded that fuzzy conceptual graphs is suitable language for knowledge representation to be used by Semantic web. Takagi and Tajima (2001) presented the conceptual matching of text notes to be used by search engines. An new search engine proposed which conceptually matches keywords and the web pages. Conceptual fuzzy set has been used for context-dependent keyword expansion. A new structure for search engine has been proposed which can resolve the context-dependent word ambiguity using fuzzy conceptual matching technique. Berenji (2001) used Fuzzy Reinforcement Learning (FRL) for text data mining and Internet search engine. Choi (2001) presented a new technique, which integrates document index with perception index. The techniques can be used for refinement of fuzzy queries on the Internet. It has been concluded that the use of perception index in commercial search engine provides a framework to handle fuzzy terms (perception-based), which is further step toward a human-friendly, natural language-based interface for the Internet. Sanchez (2001) presented the concept of Internet-based fuzzy Telerobotic for the WWW. The system receives the information from human and has the capability for fuzzy reasoning. It has be proposed to use fuzzy applets such as fuzzy logic propositions in the form of fuzzy rules that can be used for smart data base search. Bautista and Kraft (2001) presented an approach to use fuzzy logic for user profiling in Web retrieval applications. The technique can be used to expand the queries and knowledge extraction related to a group of users with common interest. Fuzzy representation of terms based on linguistic qualifiers has been used for their study. In addition, fuzzy clustering of the user profiles can be used to construct fuzzy rules and inferences in order to modify queries. The result can be used for knowledge extraction from user profiles for marketing purposes. Yager (2001) introduced fuzzy aggregation methods for intelligent search. It is concluded that the new technique can increase the expressiveness in the queries. Widyantoro and Yen (2001) proposed the use of fuzzy ontology in search engines. Fuzzy ontology of term relations can be built automatically from a collection of documents. The proposed fuzzy ontology can be used for query refinement and to suggest narrower and broader terms suggestions during user search activity. Presser (2001) introduced fuzzy logic for rule-based personalization and can be implemented for personalization of newsletters. It is concluded that the use of fuzzy logic provide better flexibility and better interpretation which helps in keeping the knowledge bases easy to maintain. Zhang et al. (2001a) presented granular fuzzy technique for web search engine to increase Internet search speed and the Internet quality of service. The techniques can be used for personalized fuzzy web search engine, the personalized granular web search agent. While current fuzzy search engines uses keywords, the proposed technique provide a framework to not only use traditional fuzzy-key-word but also fuzzy-user-preference-based search algorithm. It is concluded that the proposed model reduces web search redundancy, increase web search relevancy, and decrease user's web search time. Zhang et al. (2001b) proposed fuzzy neural web agents based on granular neural network, which discovers fuzzy rules for stock prediction. Fuzzy logic can be used for web mining. Pal et al. (2002) presented issues related to web mining using soft computing framework. The main tasks of web mining based on fuzzy logic include information retrieval and generalization. Krisnapuram et al. (1999) used fuzzy c medoids and triimed medoids for clustering of web documents. Joshi and Krisnapuram (1998) used fuzzy clustering for web log data mining. Sharestani (2001) presented the use of fuzzy logic for network intruder detection. It is concluded that fuzzy logic can be used for approximate reasoning and handling detection of intruders through approximate matching; fuzzy rule and summarizing the audit log data. Serrano (2001) presented a web-based intelligent assistance. The model is an agent-based system which uses a knowledge-based model of the e-business that provide advise to user through intelligent reasoning and dialogue evolution. The main advantage of this system is based on the human-computer understanding and expression capabilities, which generate the right information in the right time.

4 Perception-Based Information Processing for Internet

One of the problems that Internet users are facing today is to find the desired information correctly and effectively in an environment that the available information, the repositories of information, indexing, and tools are all dynamic. Even though some tools were developed for a dynamic environment, they are suffering from "too much" or " too little" information retrieval. Some tools return too few resources and some tool returns too many resources *(Figure 3.).*

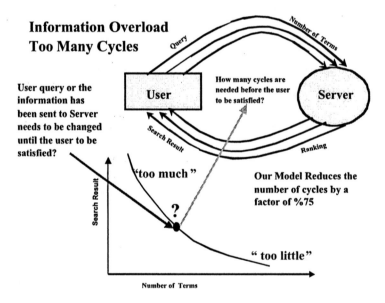

Figure 3. Information overload

The main problem with conventional information retrieval and search such as vector space representation of term-document vectors are that 1) there is no real theoretical basis for the assumption of a term and document space and 2) terms and documents are not really orthogonal dimensions. These techniques are used more for visualization and most similarity measures work about the same regardless of model. In addition, terms are not independent of all other terms. With regards to probabilistic models, important indicators of relevance may not be term -- though terms only are usually used. Regarding Boolean model, complex query syntax is often misunderstood and problems of null output and Information overload exist. One solution to these problems is to use extended Boolean model or fuzzy logic. In this case, one can add a fuzzy quantifier to each term or concept. In addition, one can interpret the AND as fuzzy-MIN and OR as fuzzy-MAX functions. Alternatively, one can add agents in the user interface and assign certain tasks to them or use machine learning to learn user behavior or preferences to improve performance. This technique is useful when past behavior is a useful predictor of the future and wide variety of behaviors amongst users exist.

5 Fuzzy Conceptual Model and Search Engine

One can use clarification dialog, user profile, context, and ontology, into a integrated frame work to address some of the issues related to search engines were described earlier. In our perspective, we define this framework as *Fuzzy Conceptual Matching based on Human Mental Model.* The Conceptual Fuzzy Set (CFS) model will be used for intelligent information and knowledge retrieval through conceptual matching of both text and images (here defined as "Concept").

5.1 Search Engine based on Conceptual Matching of Text Notes

First, a fuzzy set is defined by enumerating its elements and the degree of membership of each element. It is useful for retrieving information which includes not only the keyword, but also elements of the fuzzy set labeled by the input keyword. For example, a search engine may use baseball, diving, skiing, etc., as kinds of sports, when a user inputs "sports" as the keyword.

Second, the same word can have various meanings. Several words are used concurrently in usual sentences, but each word has multiple possible meanings (region), so we suppose an appropriate context which suits all regions of meaning of all words (**Figure 4**). At the same time, the context determines the meaning of each word.

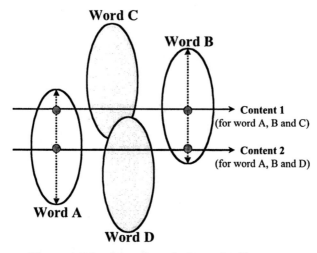

Figure 4. Meanings of words determined by a context.

For example, "sports" may mean "diving" or "sailing" when it is used with "marine," and may mean "baseball" or "basketball" when used with "TV programs." That is, each possibility distribution of meaning is considered as a fuzzy set itself. For information retrieval, keyword expansion that considers context is necessary, because simple expansion of a possible region causes a flood of words. For example, even if the user intends "marine sports," the set of expanded keywords includes unnecessary sports such as "baseball." However, an ordinary fuzzy set does not provide us the method to deal with context-dependent word ambiguity. To overcome this problem, we previously proposed using conceptual fuzzy sets (CFSs) (Takagi et al. 1995, 1996, 1999a and 1999b), which conform to Wittgenstein's concept, to represent the meanings of concepts.

5.1.2 Fuzzy Sets and Context Dependent Word Ambiguity

We propose a search engine which conceptually matches input keywords and web pages. The conceptual matching is realized by context-dependent keyword expansion using conceptual fuzzy sets.

5.1.2.1 Conceptual Fuzzy Sets (Takagi et al. 1995, 1996, 1999a and 1999b)

Let's think about the meaning of "heavy." A person weighting 100kg would usually be considered heavy. But there is no clear boundary between "heavy" and "not heavy." Fuzzy sets are generally used to indicate these regions. That is, we have a problem of specificity.

Figure 5. The meaning of "heavy."

Let's think about it some more. For a vehicle, "heavy" might be several thousand kgs. For a ship, it might be more than ten thousand tons. Therefore, the item "heavy" being judged affects the vagueness of the meaning of "heavy" much more than the non-specificity of the amount when the item is already determined as shown in **Figure 5.** Moreover, "work" can be heavy, "traffic" can be heavy, and "smoking" can be heavy. So the meaning of "heavy" changes with the context, which results in the vagueness of the meaning.

That is, the main cause of vagueness is ambiguity in the language, not specificity. Ordinary fuzzy set theory has not dealt with the context dependent meaning representation concerning language ambiguity. However, as we mentioned in the Introduction, a fuzzy set is defined by enumerating its elements and the degree of membership of each element, we can use it to express word ambiguity by enumerating all possible meanings of a word, then estimating the degrees of compatibilities between the word and the meanings. Fuzzy set theory should therefore deal with language ambiguity as the main cause of vagueness.

To overcome this problem, we previously proposed using conceptual fuzzy sets. Although several works have been published, we will explain CFSs for understanding the following section. According to Wittgenstein (1953), the meaning of a concept can be represented by the totality of its uses. In this spirit, conceptual fuzzy sets, in which a concept is represented by the distribution of the activation concepts, are proposed.

The label of a fuzzy set represents the name of a concept, and a fuzzy set represents the meaning of the concept. Therefore, the shape of a fuzzy set is determined by the meaning of the label, which depends on the situation (**Figure 6**).

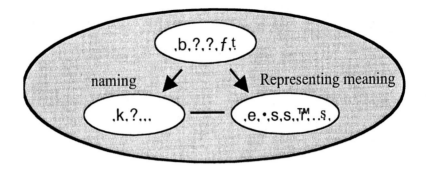

Figure 6. A fuzzy set as a meaning representation.

According to the theory of "meaning representation from use" proposed by Wittgenstein) the various meanings of a label (word) can be represented by other labels (words), and we can assign grades of activation showing the degree of compatibility between labels.

A CFS achieves this by using distributions of activations. A CFS is realized as an associative memory in which a node represents a concept and a link represents the strength of the relation between two (connected) concepts. The activation values agreeing with the grades of membership are determined through this associative memory. In a CFS, the meaning of a concept is represented by the distribution of the activation values of the other nodes. The distribution evolves from the activation of the node representing the concept of interest. The image of a CFS is shown in **Figure 7**.

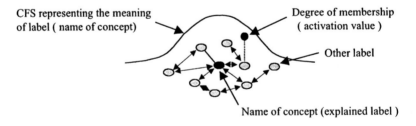

Figure 7. A conceptual fuzzy set represented by associative memories.

5.1.2.2 CFS Representing a Composed Concept having Multiple Meanings Depending on the Situation

Because the distribution changes depending on which labels are activated as a result of the conditions the activations show a context-dependent meaning. When more than two labels are activated, a CFS is realized by the overlapping propagations of their activations. In CFS notation, operations and their controls are all realized by the distribution of the activations and their propagations in the associative memories.

We can say that the distribution determined by the activation of a label agrees with the region of thought corresponding to the word expressing its meaning. The distribution (meaning of a label), that is a figure of a fuzzy set, changes depending on considered aspects that reflect the context.

5.1.3 Creating of Conceptual Fuzzy Sets

Previously we used bidirectional associative memories (BAMs) Kasko 1987 and 1992) to generate CFSs, because of the clarity of the constraints used for their utilization. In this paper, we use Hopfield Networks, whose output can be also used with a continuous value, to overcome the limitation of BAMs that are a layered neural network. We do so because in a correlative neural network, relationships between concepts may not be a layered structure.

The following shows how to construct CFSs using Hopfield Networks (Hopfield 1982 and 1984).

Memorizing pieces of knowledge:
1. Classify piece of knowledge into several aspects. One piece becomes one pattern and one aspect corresponds to one Hopfield Network.
2. Each Hopfield network contains multiple patterns in the same aspect.

Generating CFSs:
1. Recollect patterns which include a node of interest in each Hopfield Network.
2. Sum all recollected patterns and normalize the activation values.

Figure 5 shows the image of memorized patterns and a generated CFS.

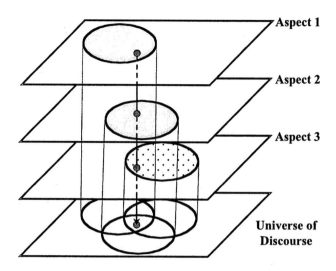

Figure 8. Image of memorized patterns and a generated CFS.

Logic-based paradigms for knowledge representation use symbolic processing both for concept representation and inference. Their underlying assumption is that a concept can be defined precisely. This causes an exponential increase in the number of cases, because we have to memorize all cases according to every possible context. In contrast, knowledge representation using CFSs memorizes knowledge about generalizations instead of exact cases (**Figure 8**). The following is an example to compare the proposed knowledge representation with ordinary logic-based paradigms. It shows that context-dependent meanings are generated by activating several keywords.

5.1.4 Conceptual Matching in a Search Engine Using CFS

Information retrieval in the Internet is generally done by using keyword matching, which requires that for words to match, they must be the same or synonyms. But essentially, not only the information that matches the keywords exactly, but also information related in meaning to the input keywords should be retrieved. The following reasons are why fuzzy sets are essential for information retrieval.

5.1.4.1 Scheme of Search Engine (Kobayashi and Takeda 2000, Quarino and Vetere 1999)

Usually, search engines work as follows.

Index collecting of Web pages:
An indexer extracts words from Web pages, analyzes them, and stores the result as indexing information.
Retrieving information:
The Web pages, which include input keywords, are extracted. The pages are assigned priority and are sorted referencing the indexing information above.

As we mentioned in the Introduction, information retrieval is generally done by using keyword matching, which requires words to match and is different from conceptual matching.

5.1.4.2 Conceptual Matching

We propose a search engine system which conceptually matches input keywords and Web pages according to the following idea.

1.   Expand input keywords to the elements of CFSs.
2.   Evaluate the matching degree between the set of expanded keywords and the set of words included in each Web page.
3.   Sort the Web pages and display them according to the matching degrees.

The following shows the process in the proposed search engine.

Index collecting of Web pages:
1.   Extract nouns and adjectives, and count the frequency of each word.
2.   Calculate an evaluation of each word using the TF-IDF method for each Web page.
3.   Store the evaluation into a lexicon.

Retrieve information:

1.   A user inputs keywords into a browser, which transfers the keywords to a CFS unit.
2.   Propagation of activation occurs from input keywords in the CFS unit. The meanings of the keywords are represented in other expanded words using conceptual fuzzy sets, and the activation value of each word is stored into the lexicon.
3.   Matching is executed in the following process for each Web page. Obtain

the final evaluation of each word by multiplying the evaluation by the TF-IDF method and the activation value. Sum up the final evaluations of all words and attach the result to each Web page as a matching degree.

4. The matched Web pages are sorted according to the matching degrees, and their addresses are returned to the browser with their matching degrees.

5.1.5 Simulations and Evaluations

Let's think about the case where we are searching for Web pages of places to visit using certain keywords, we indexed 200 actual Web pages, and compared the search result of the following two matching methods.

1. TF-IDF method
2. our proposed method (**Figure 9** using CFS)

Evaluation 1:

If the CFS unit has knowledge in fuzzy sets about places, and if a user inputs "famous resort" as a keyword, relating name of places are added as expanded keywords with their activation degrees agreeing with membership degrees.

Famous resort = 0.95/gold coast + 0.95/the Cote d'Azur + 0.91/Fiji + ..

Table 6 shows the result when "famous resort" and "the Mediterranean Sea" are input as keywords. It consists of names of places and activation values, and shows the extended keywords generated by the activation of the above two keywords.

Table 6. Extended keywords

Ranking	Word	Activation Value
,P	The Côte d'Azur	1.0000
2	The Mediterranean Sea	0.9773
3	Famous Resort	0.9187
,S	Crete	0.8482
5	Capri	0.6445
6	Anguilla	0.6445
7	Santorini	0.6445
8	Taormina	0.6445
9	Sicily	0.4802
10	Gold Coast	0.0748

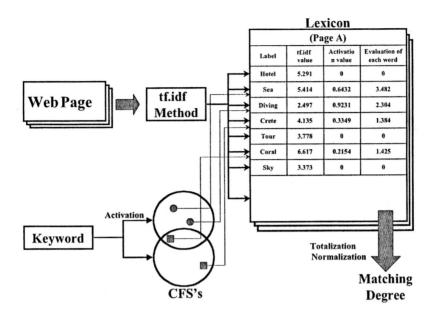

Figure 9. Scheme of the proposed search engine.

Evaluation 2:

If the CFS unit memorizes knowledge about "vacation" and "sports" such as,

vacation = 1.0/vacance + 0.6/sea + 0.6/sandy beach + 0.6/the South Pacific + ..
sports = 1.0/spots + 0.6/diving + 0.6/trekking + 0.6/golf + ..

then a ranked list of Web pages appears. Table 9 shows the extended keywords generated by the activation of "vacation" and "sports."

From the results, we demonstrate the effectiveness of our proposed method. Unlike the first case (**Table 7**), the Web pages were not retrieved by place names, but by the activities corresponding to the context of "vacation sports." Jet skiing and surfing were suggested by the CFS as a relevant sports, but baseball was not.

We show that pertinent Web pages can be retrieved independently of the key ward, because even though the region "sport" can encompass a huge number of different activities.

Table 7. Extended keywords.

Ranking	Word	Activation Value
,P	Diving	0.6079
1	Surfing	0.6079
3	Sports	0.5000
3	Vacation	0.5000
5	Golf	0.3039
5	Rock Climbing	0.3039
5	Baseball	0.3039
5	Sea	0.3039
5	Paradise	0.3039
5	SandyBearch	0.3039

6 Challenges and Road Ahead

During the August 2001, BISC program hosted a workshop toward better u   n-
derstanding of the issues related to the Internet (Fuzzy Logic and the Internet-
FLINT2001, Toward the Enhancing the Power of the Internet). The main purpose
of the Workshop was to draw the attention of the fuzzy logic community as well
as the Internet community to the fundamental importance of specific Internet-
related problems. This issue is critically significant about problems that center on
search and deduction in large, unstructured knowledge bases. The Workshop provided a unique opportunity for the academic and corporate communities to address
new challenges, share solutions, and discuss research directions for the future.
Following are the areas that were recognized as challenging problems and the new
direction toward the next generation of the search engines and Internet. We summarize the challenges and the road ahead into four categories as follows:

1. *Search Engine and Queries:*

- Deductive Capabilities
- Customization and Specialization
- Metadata and Profiling
- Semantic Web
- Imprecise-Querying
- Automatic Parallelism via Database Technology

- Approximate Reasoning
- Ontology
- *Ambiguity Resolution through Clarification Dialog; Definition/Meaning & Specificity*User Friendly
- Multimedia
- Databases
- Interaction

2. *Internet and the Academia:*

- Ambiguity and Conceptual and Ontology
- Aggregation and Imprecision Query
- Meaning and structure Understanding
- Dynamic Knowledge
- Perception, Emotion, and Intelligent Behavior
- Content-Based
- Escape from Vector SpaceDeductive Capabilities
- Imprecise-Querying
- *Ambiguity Resolution through Clarification Dialog*
- *Precisiated Natural Languages (PNL)*

3. *Internet and the Industry:*

- XML=>Semantic Web
- Workflow
- Mobile E-Commerce
- CRM
- Resource Allocation
- Intent
- Ambiguity Resolution
- Interaction
- Reliability
- Monitoring
- Personalization and Navigation
- Decision Support
- Document Soul
- Approximate Reasoning
- Imprecise Query
- Contextual Categorization

4. *Fuzzy Logic and Internet; Fundamental Research:*

- Computing with Words (CW)
- Computational Theory of Perception (CTP)
- Precisiated Natural Languages (PNL)

7 Conclusion

Intelligent search engines with growing complexity and technological challenges are currently being developed. This requires new technology in terms of understanding, development, engineering design and visualization. While the technological expertise of each component becomes increasingly complex, there is a need for better integration of each component into a global model adequately capturing the imprecision and deduction capabilities. In addition, intelligent models can mine the Internet to conceptually match and rank homepages based on predefined linguistic formulations and rules defined by experts or based on a set of known homepages.

We also showed the necessity and also the problems of applying fuzzy sets to information retrieval using FCM Framework. Next, we introduced using conceptual fuzzy sets in overcoming those problems, and proposed the realization of conceptual fuzzy sets using Hopfield Networks. Based on above, we proposed the architecture of the search engine which can execute conceptual matching dealing with context-dependent word ambiguity. Finally, we evaluated our proposed method through two simulations of retrieving actual web pages, and compare the proposed method with the ordinary TF-IDF method. We showed that our method could correlate seemingly unrelated input keywords and produce matching Web pages, whereas the simple TF-IDF method could not.

The FCM model can be used as a framework for intelligent information and knowledge retrieval through conceptual matching of both text and images (here defined as "Concept"). The FCM can also be used for constructing fuzzy ontology or terms related to th context of the query and search to resolve the ambiguity. This model can be used to calculate conceptually the degree of match to the object or query.

Acknowledgement

Funding for this research was provided by the British Telecommunication (BT) and the BISC Program of UC Berkeley.

References

1. J. Baldwin, Future directions for fuzzy theory with applications to intelligent agents, in M. Nikravesh and B. Azvine, FLINT 2001, New Directions in Enhancing the Power of the Internet, UC Berkeley Electronics Research Laboratory, Memorandum No. UCB/ERL M01/28, August 200.
2. J. F. Baldwin and S. K. Morton, conceptual Graphs and Fuzzy Qualifiers in Natural Languages Interfaces, 1985, University of Bristol.
3. M. J. M. Batista et al., User Profiles and Fuzzy Logic in Web Retrieval, in M. Nikravesh and B. Azvine, FLINT 2001, New Directions in Enhancing the Power of the Internet, UC Berkeley Electronics Research Laboratory, Memorandum No. UCB/ERL M01/28, August 2001.
4. H. Beremji, Fuzzy Reinforcement Learning and the Internet with Applications in Power Management or wireless Networks, in M. Nikravesh and B. Azvine, FLINT 2001, New Directions in Enhancing the Power of the Internet, UC Berkeley Electronics Research Laboratory, Memorandum No. UCB/ERL M01/28, August 2001.
5. T.H. Cao, Fuzzy Conceptual Graphs for the Semantic Web, in M. Nikravesh and B. Azvine, FLINT 2001, New Directions in Enhancing the Power of the Internet, UC Berkeley Electronics Research Laboratory, Memorandum No. UCB/ERL M01/28, August 2001.
6. D. Y. Choi, Integration of Document Index with Perception Index and Its Application to Fuzzy Query on the Internet, in M. Nikravesh and B. Azvine, FLINT 2001, New Directions in Enhancing the Power of the Internet, UC Berkeley Electronics Research Laboratory, Memorandum No. UCB/ERL M01/28, August 2001.
7. N. Guarino, C. Masalo, G. Vetere, "OntoSeek : content-based access to the Web", IEEE Intelligent Systems, Vol.14, pp.70-80 (1999)
8. K.H.L. Ho, Learning Fuzzy Concepts by Example with Fuzzy Conceptual Graphs. In 1st Australian Conceptual Structures Workshop, 1994. Armidale, Australia.
9. J. J. Hopfield, "Neural networks and physical systems with emergent collective computational abilities", Proceedings of the National Academy of Sciences U.S.A., Vol.79, pp.2554-2558 (1982)
10. J. J. Hopfield, "Neurons with graded response have collective computational properties like those of two-state neurons, Proceedings of the National Academy of Sciences U.S.A., Vol.81, pp.3088-3092 (1984)
11. A. Joshi and R. Krishnapuram, Robust Fuzzy Clustering Methods to Support Web Mining, in Proc Workshop in Data Mining and Knowledge Discovery, SIGMOD, pp. 15-1 to 15-8, 1998.
12. M. Kobayashi, K. Takeda, "Information retrieval on the web", ACM Computing Survey, Vol.32, pp.144-173 (2000)
13. B. Kosko, "Adaptive Bi-directional Associative Memories," Applied Optics, Vol. 26, No. 23, 4947-4960 (1987).
14. B. Kosko, "Neural Network and Fuzzy Systems," Prentice Hall (1992).
15. R. Krishnapuram et al., A Fuzzy Relative of the K-medoids Algorithm with application to document and Snippet Clustering , in Proceedings of IEEE Intel. Conf. Fuzzy Systems-FUZZIEEE 99, Korea, 1999.
16. T. B. Lee, Transcript of Tim Berners-Lee's talk to the LCS 35th Anniversary celebrations, Cambridge Massachusetts, 1999/April/14

17. D. B. Lenat, From 2001 to 2001: Common Sense and the Mind of HAL; A chapter from Hal's Legacy: 2001 as Dream and Reality (http://www.cyc.com/publications.html)
18. T. P. Martin, Searching and smushing on the Semantic Web – Challenges for Soft Computing, in M. Nikravesh and B. Azvine, FLINT 2001, New Directions in Enhancing the Power of the Internet, UC Berkeley Electronics Research Laboratory, Memorandum No. UCB/ERL M01/28, August 2001.
19. M. Nikravesh and B. Azvine, FLINT 2001, New Directions in Enhancing the Power of the Internet, UC Berkeley Electronics Research Laboratory, Memorandum No. UCB/ERL M01/28, August 2001.
20. M. Nikravesh, Fuzzy Logic and Internet: Perception Based Information Processing and Retrieval, Berkeley Initiative in Soft Computing, Report No. 2001-2-SI-BT, September 2001a.
21. M. Nikravesh, BISC and The New Millennium, Perception-based Information Processing, Berkeley Initiative in Soft Computing, Report No. 2001-1-SI, September 2001b.
22. S. K. Pal, V. Talwar, and P. Mitra, Web Mining in Soft Computing Framework: Relevance, State of the Art and Future Directions, to be published in IEEE Transcations on Neural Networks, 2002.
23. G. Presser, Fuzzy Personalization, in M. Nikravesh and B. Azvine, FLINT 2001, New Directions in Enhancing the Power of the Internet, UC Berkeley Electronics Research Laboratory, Memorandum No. UCB/ERL M01/28, August 2001.
24. E. Sanchez, Fuzzy logic e-motion, in M. Nikravesh and B. Azvine, FLINT 2001, New Directions in Enhancing the Power of the Internet, UC Berkeley Electronics Research Laboratory, Memorandum No. UCB/ERL M01/28, August 2001.
25. A. M. G. Serrano, Dialogue-based Approach to Intelligent Assistance on the Web, in M. Nikravesh and B. Azvine, FLINT 2001, New Directions in Enhancing the Power of the Internet, UC Berkeley Electronics Research Laboratory, Memorandum No. UCB/ERL M01/28, August 2001.
26. S. Shahrestani, Fuzzy Logic and Network Intrusion Detection, in M. Nikravesh and B. Azvine, FLINT 2001, New Directions in Enhancing the Power of the Internet, UC Berkeley Electronics Research Laboratory, Memorandum No. UCB/ERL M01/28, August 2001.
27. T. Takagi and M. Tajima, Proposal of a Search Engine based on Conceptual Matching of Text Notes, in M. Nikravesh and B. Azvine, FLINT 2001, New Directions in Enhancing the Power of the Internet, UC Berkeley Electronics Research Laboratory, Memorandum No. UCB/ERL M01/28, August 2001.
28. T. Takagi, A. Imura, H. Ushida, and T. Yamaguchi, "Conceptual Fuzzy Sets as a Meaning Representation and their Inductive Construction," International Journal of Intelligent Systems, Vol. 10, 929-945 (1995).
29. T. Takagi, A. Imura, H. Ushida, and T. Yamaguchi, "Multilayered Reasoning by Means of Conceptual Fuzzy Sets," International Journal of Intelligent Systems, Vol. 11, 97-111 (1996).
30. T. Takagi, S. Kasuya, M. Mukaidono, T. Yamaguchi, and T. Kokubo, "Realization of Sound-scape Agent by the Fusion of Conceptual Fuzzy Sets and Ontology," 8th International Conference on Fuzzy Systems FUZZ-IEEE'99, II, 801-806 (1999).

31. T. Takagi, S. Kasuya, M. Mukaidono, and T. Yamaguchi, "Conceptual Matching and its Applications to Selection of TV Programs and BGMs," IEEE International Conference on Systems, Man, and Cybernetics SMC'99, III, 269-273 (1999).

32. Wittgenstein, "Philosophical Investigations," Basil Blackwell, Oxford (1953).

33. R. Yager, Aggregation Methods for Intelligent Search and Information Fusion, in M. Nikravesh and B. Azvine, FLINT 2001, New Directions in Enhancing the Power of the Internet, UC Berkeley Electronics Research Laboratory, Memorandum No. UCB/ERL M01/28, August 2001.

34. John Yen, Incorporating Fuzzy Ontology of Terms Relations in a Search Engine, in M. Nikravesh and B. Azvine, FLINT 2001, New Directions in Enhancing the Power of the Internet, UC Berkeley Electronics Research Laboratory, Memorandum No. UCB/ERL M01/28, August 2001.

35. L. A.Zadeh, The problem of deduction in an environment of imprecision, uncertainty, and partial truth, in M. Nikravesh and B. Azvine, FLINT 2001, New Directions in Enhancing the Power of the Internet, UC Berkeley Electronics Research Laboratory, Memorandum No. UCB/ERL M01/28, August 2001 [2001a].

36. L.A. Zadeh, A Prototype-Centered Approach to Adding Deduction Capability to Search Engines -- The Concept of Protoform, BISC Seminar, Feb 7, 2002, UC Berkeley, 2002.

37. L. A. Zadeh, " A new direction in AI – Toward a computational theory of perceptions, AI Magazine 22(1): Spring 2001, 73-84

38. L.A. Zadeh, From Computing with Numbers to Computing with Words-From Manipulation of Measurements to Manipulation of Perceptions, IEEE Trans. On Circuit and Systems-I Fundamental Theory and Applications, 45(1), Jan 1999, 105-119.

39. Y. Zhang et al., Granular Fuzzy Web Search Agents, in M. Nikravesh and B. Azvine, FLINT 2001, New Directions in Enhancing the Power of the Internet, UC Berkeley Electronics Research Laboratory, Memorandum No. UCB/ERL M01/28, August 2001.

40. Y. Zhang et al., Fuzzy Neural Web Agents for Stock Prediction, in M. Nikravesh and B. Azvine, FLINT 2001, New Directions in Enhancing the Power of the Internet, UC Berkeley Electronics Research Laboratory, Memorandum No. UCB/ERL M01/28, August 2001.

Fuzzy Logic and Rank Aggregation for the World Wide Web

M. M. Sufyan Beg and Nesar Ahmad

Department of Electrical Engineering, Indian Institute of Technology, New Delhi, 110 016, India, {mmsbeg,nahmad}@ee.iitd.ac.in

Abstract

Rank Aggregation is the problem of generating a near-"consensus" ranking for a given set of rankings. In literature, fuzzy logic has been extensively employed to reach consensus in the domain of group decision making. Therefore, we feel that the application of fuzzy theory to the problem of rank aggregation, which may broadly be taken as a special case of group decision making, should lead to improvement in achieving a higher degree of consensus. It is for this reason that the adoption of fuzzy techniques is being investigated in this chapter for the problem of rank aggregation on the web. When applied to the World Wide Web, this work finds applications in meta-searching, search engine comparison, spam fighting and word association techniques. Classical fuzzy rank ordering techniques have been adopted for web applications. A few novel techniques for rank aggregation that outshine the existing techniques, are also proposed.

Introduction

In an Internet search, the user uses a query language to describe the nature of documents, and, in response, a search engine locates the documents that "best match" the description. A number of search engines are available to Internet users today and more are likely to appear in the future. These systems differ from one another in the indexing technique they use to construct the repository of documents and the search algorithm they employ to generate the response. As a result, the results for the same query from different search engines vary considerably. There is a need, therefore, to get the "collective wisdom" of different public web search engines, by aggregating the ranked results returned by these search engines.

Rank aggregation is the problem of collating a given set of rankings. In applications like sports, rank aggregation may be used to declare the overall team positions based on the rankings given by various judges. When applied to the web, this finds a primary application in meta-searching the World Wide Web. A meta-

27

search engine is the one that doesn't have a database of its own, rather it takes the search results from other public search engines, collate those results and present the combined result before the user. This is how we get the combined advantage of different search techniques being employed by the participating search engines.

Another usage of rank aggregation is in spam fighting. The commercial interest of many web page authors to have their pages ranking high for certain queries, forces them to do some illegitimate manipulations. These manipulations may go as far as adding text in an invisible ink. This is clearly an undesirable situation. If a page spams fewer than half the participating search engines, the spam page would occupy the bottom partition of that aggregated ranking which satisfies *Condorcet* criterion (see definition 9). However, for a page spamming more than half the search engines, not much can be done anyway --- garbage in, garbage out.

One more novel application of rank aggregation to the web has been brought to light in (Dwork et al 2001). This application appears as what is called as *multicriteria selection* and *word association queries*. An average web searcher may not be very good at formulating queries with precise AND and OR operators. These are the situations where the user knows a list of keywords that collectively describe the topic, but doesn't know precisely whether to use AND operators or the OR operators to combine these keywords. It may be noted that the AND operator requiring *all* the keywords to be present in the document would return too few documents, while the OR operator requiring *any* of the keywords in the document would return too many documents. To ease out such a situation, the keywords are taken from the user and several sub-queries with all subsets of these keywords are formulated using AND operator. The results to all these sub-queries are then aggregated to get the final ranking.

In this chapter, we describe various methods to carry out rank aggregation of the list of documents obtained from different search engines in response to a query. We have adopted a few existing fuzzy rank ordering techniques and also proposed some other novel techniques.

Background and Related Work

Let us begin with some useful definitions, illustrated with appropriate examples.

Definition 1. Given a universe U and $T \subseteq U$, an *ordered list* (or simply, a *list*) l with respect to U is given as $l = [d_1, d_2, ..., d_{|T|}]$, with each $d_i \in T$, and $d_1 \succ d_2 \succ ... \succ d_{|T|}$, where "$\succ$" is some ordering relation on T. Also, for $i \in U \wedge i \in l$, let $l(i)$ denote the position or rank of i, with a higher rank having a lower numbered position in the list. We may assign a unique identifier to each element in U, and thus, without loss of generality, we may get $U = \{1, 2, ..., |U|\}$.

Definition 2. Full List: If a list l contains all the elements in U, then it is said to be a *full list*.

Example 1. A full list l given as $[c,d,b,a,e]$ has the ordering relation $c \succ d \succ b \succ a \succ e$. The universe U may be taken as $\{1,2,3,4,5\}$ with, say, $a \equiv 1$, $b \equiv 2$, $c \equiv 3$, $d \equiv 4$ and $e \equiv 5$. With such an assumption, we have $l = [3,4,2,1,5]$. Here $l(3) \equiv l(c)=1$, $l(4) \equiv l(d)=2$, $l(2) \equiv l(b)=3$, $l(1) \equiv l(a)=4$, $l(5) \equiv l(e)=5$.

Given two lists l_1 and l_2, measures of similarity between the two lists are given as definitions 3 and 4. There are many more measures, but we shall primarily concentrate on these two in this chapter.

Definition 3. Kendall Tau distance: The *Kendall tau distance* between two full lists l_1 and l_2, each of cardinality $|l|$, is given as follows.

$$K(l_1,l_2) = \frac{\left| \{(i,j) \mid \forall l_1(i) < l_1(j), l_2(i) > l_2(j)\} \right|}{(1/2)\, |l|\, (\, |l|-1)}$$

Definition 4. Spearman footrule distance: The *Spearman footrule distance* (SFD) between two full lists l_1 and l_2, each of cardinality $|l|$, is given as follows.

$$F(l_1,l_2) = \frac{\sum_{\forall i} |l_1(i) - l_2(i)|}{\left\lfloor \left(\frac{1}{2} \right) |l|^2 \right\rfloor}$$

Example 2. For $l_1 = [c,d,b,a,e]$, $l_2 = [b,d,e,c,a]$, the Spearman footrule distance may be obtained as follows.

$$F(l_1,l_2) = \frac{|4-5| + |3-1| + |1-4| + |2-2| + |5-3|}{\lfloor 0.5 \times 5^2 \rfloor} = \frac{8}{12} = 0.667$$

Definition 5. Given a set of k full lists as $L=\{l_1, l_2,\dots, l_k\}$, the *normalized aggregated Kendall distance* of a full list l to the set of full lists L is given as

$$K(l,L) = \frac{\sum_{i=1}^{k} K(l,l_i)}{k}$$, while the *normalized aggregated footrule distance* of l to L

is given as $F(l,L) = \dfrac{\sum_{i=1}^{k} F(l,l_i)}{k}$.

Definition 6. Rank Aggregation: Given a set of lists $L=\{l_1, l_2,\dots, l_k\}$, *Rank Aggregation* is the task of coming up with a list l such that either $K(l,L)$ or $F(l,L)$ is minimized.

The rank aggregation obtained by optimizing the Kendall distance is called *Kemeny optimal aggregation* (KOA), and it has been shown in (Dwork et al 2001) that KOA is NP-hard even when k=4.

Proposition 1 (Diaconis and Graham 1977)**:** For any two full lists l_1 and l_2,
$$K(l_1, l_2) < F(l_1, l_2) < 2K(l_1, l_2)$$

Based on Proposition 1, Kendall distance is approximated by the Spearman footrule distance.

Proposition 2 (Dwork et al 2001): If l is the Kemeny optimal aggregation of the set of full lists $L=\{l_1, l_2,..., l_k\}$ and l' optimizes the footrule aggregation, then
$$K(l', L) < 2K(l, L).$$

A polynomial time algorithm has been given in (Dwork et al 2001) to compute the footrule optimal aggregation (FOA) for full lists. However, when collating the results from various search engines, the lists are almost invariably always the partial ones.

Definition 7. Partial List: A list l containing elements, which are a strict subset of U, is called a partial list. We have a strict inequality $|l|<|U|$.

The FOA for partial lists is stated in (Dwork et al 2001) to be equivalent to the NP-hard problem of computing the number of edges to delete to convert a directed graph into a *directed acyclic graph* (DAG). Let us call the problem of computing FOA for partial lists as *partial footrule optimal aggregation* (PFOA) problem. The NP-hard nature of PFOA forces us to go back to some "positional" methods, in which each candidate is assigned a score based on the position in which it appears in the list of each voter, and then the candidates are sorted based on their total score. A classical positional method is the one given by J. C. Borda in (Borda 1781).

Definition 8. Borda's Method (BM) of Rank Aggregation: Given k lists $l_1, l_2,..., l_k$, for each candidate c_j in list l_i, we assign a score $S_i(c_j) = |c_p: l_i(c_p) > l_i(c_j)|$. The candidates are then sorted in a decreasing order of the total Borda score

$$S(c_j) = \sum_{i=1}^{k} S_i(c_j).$$

Example 3. Given lists $l_1 = [c,d,b,a,e]$ and $l_2 = [b,d,e,c,a]$.

$S_1(a)=|e|=1$, as $l_1(e)=5 > l_1(a)=4$.

Similarly,

$S_1(b)=|a,e|=2$, as $l_1(e)=5 > l_1(b)=3$ and $l_1(a)=4 > l_1(b)=3$.

Proceeding this way, we get

$S_1(c) = |a,b,d,e| = 4,$
$S_1(d) = |a,b,e| = 3,$
$S_1(e) = || = 0,$
$S_2(a) = || = 0,$
$S_2(b) = |a,c,d,e| = 4,$
$S_2(c) = |a| = 1,$
$S_2(d) = |a,c,e| = 3,$
$S_2(e) = |a,c| = 2,$
$S(a) = S_1(a)+S_2(a) = 1+0 = 1,$
$S(b) = S_1(b)+S_2(b) = 2+4 = 6,$
$S(c) = S_1(c)+S_2(c) = 4+1 = 5,$
$S(d) = S_1(d)+S_2(d) = 3+3 = 6,$
$S(e) = S_1(e)+S_2(e) = 0+2 = 2.$

Now, sorting the elements based on their total scores, we get the combined ranking as $b \approx d \succ c \succ e \succ a$. The '$\approx$' symbol indicates a tie.

Definition 9. Condorcet Criteria: If there is some element $d_i \in T$ (where $T \subseteq U$), which defeats every other element in pairwise simple majority voting, then this element should be ranked first.

The positional methods are linear in complexity, but the problems with them are that they neither optimize any *distance* criterion nor satisfy the *Condorcet* property, which is so very essential for spam fighting (Dwork et al 2001). These problems of the positional methods on one hand, and the NP-hard nature of PFOA on the other, motivated us (Beg and Ahmad 2002) to apply *genetic algorithm* for the PFOA problem. But the successive generations of GA take increasing amount of time. Since, fuzzy logic has been extensively studied in literature for arriving at consensus in group decision making, the adoption of some fuzzy techniques is being investigated in this chapter for getting an improvement over the Borda's method.

Fuzzy Methods for Rank Aggregation

The strong statement in (Ross 1997) that reads as "... but there certainly should be no restrictions on the usefulness of fuzzy information in the *process* of making a decision or of coming to some consensus", has served as a motivation for the work being discussed in this section. Since our goal in this chapter has been to reach as best a consensus amongst the participating search engines as possible, fuzzy logic is worth giving a try towards this goal (Ahmad and Beg 2002). First, some classical fuzzy ranking techniques are adopted for the problem of rank aggregation. Taking lesson out of their comparison studies, a few new heuristics, namely, MFO, MBV, Improved Shimura and Entropy-based are then proposed.

The objective function to be minimized is the sum of Spearman footrule distances between the aggregated rank l and each of the N participating partial lists. Let us assume that the set of documents constituting the search result from the ith search engine is $R_i = \left\{ d_1^i, d_2^i, ..., d_{n_i}^i \right\}$. Let the union of all such sets from the N participating engines be $U = \bigcup_{i=1}^{N} R_i$. Without loss of generality, we can assume that $U = \left\{ 1, 2, ..., |U| \right\}$.

Let there be a partial list l_j and a full list l, with the number of elements in them being $|l_j|$ and $|U|$, respectively. In order to evaluate the Spearman footrule distance between l_j and l, we first complete l_j into a full list as follows.

$$l_j\big[l_j(i)\big] = \begin{cases} unchanged, & if \ \ i \le |l_j| \\ \\ x\big|x \in U \vee x \notin l_j, & otherwise \end{cases}$$

Next we modify the positions in the full list l as follows.

31

$$l(i) = \begin{cases} unchanged, & if \ \ i \le |l_j| \\ \dfrac{\sum\limits_{k=(|l_j|+1)}^{(|U|)} l(k)}{(|U|-|l_j|)}, & otherwise \end{cases}$$

Now, we can find the Spearman footrule distance $F(l,l_j)$ between the lists l and l_j using definition 4.

Example 4. For $|U|=5$, let the full list be $l=\{5,4,1,3,2\}$ and the partial list l_j with $|l_j|=3$ be $l_j=\{2,1,4\}$. We shall first complete l_j into a full list as $l_j=\{2,1,4,3,5\}$ and also modify the positions in l as $l=\{5,4,1,2.5,2.5\}$. Now we can find the Spearman footrule distance $F(l,l_j)$ between the lists l and l_j as

$$F(l,l_j) = \frac{|5-2|+|4-1|+|1-4|+|2.5-3|+|2.5-5|}{\lfloor 0.5 \times 5^2 \rfloor} = 1.$$

With the set of N partial list of documents as $L=\{l_1,l_2,...,l_N\}$, the normalized ag-

gregated footrule distance would be $F(l,L) = \dfrac{\sum\limits_{i=1}^{N} F(l,l_i)}{N}$. Hence, this is the objective function to be minimized using the fuzzy techniques.

Shimura technique of fuzzy ordering

We begin with the Shimura technique of fuzzy ordering (Shimura 1973), as it is well suited for non-transitive rankings. For variables x_i and x_j defined on universe X, a relativity function $f(x_i|x_j)$ is taken to be the membership of preferring x_i over x_j. This function is given as

$$f(x_i|x_j) = \frac{f_{x_j}(x_i)}{\max(f_{x_j}(x_i), f_{x_i}(x_j))}$$

where, $f_{x_j}(x_i)$ is the membership function of x_i with respect to x_j, and $f_{x_i}(x_j)$ is the membership function of x_j with respect to x_i. For $X=[x_1,x_2,...,x_n]$, $f_{x_i}(x_i)=1$. $C_i = \min_{j=1}^{n} f(x_i|x_j)$ is the membership ranking value for the ith variable. Now if a descending sort on C_i ($i=1$ to n) is carried out, the sequence of $i's$ thus obtained would constitute the aggregated rank. For the lists $l_1, l_2,..., l_N$ from the N participating search engines, we can have

$$f_{x_j}(x_i) = \frac{|k \in [1,N] \vee l_k(x_i) < l_k(x_j)|}{N}$$

Example 5. Given $l_1=[3,4,2,1]$, $l_2=[2,4,3,1]$ and $l_3=[4,2,1,3]$

$$f_{x_i}(x_j) = \begin{matrix} i \\ \downarrow \end{matrix} \begin{bmatrix} 1 & 0 & 0.33 & 0 \\ 1 & 1 & 0.67 & 0.33 \\ 0.67 & 0.33 & 1 & 0.33 \\ 1 & 0.67 & 0.67 & 1 \end{bmatrix}$$

$$f(x_i|x_j) = \begin{matrix} i \\ \downarrow \end{matrix} \begin{bmatrix} 1 & 0 & 0.5 & 0 \\ 1 & 1 & 1 & 0.5 \\ 1 & 0.5 & 1 & 0.5 \\ 1 & 1 & 1 & 1 \end{bmatrix} \Rightarrow C_i = \begin{matrix} 1 \\ 2 \\ 3 \\ 4 \end{matrix} \begin{bmatrix} 0 \\ 0.5 \\ 0.5 \\ 1 \end{bmatrix}$$

A descending sort on C_i gives the aggregated rank as either $l=[4,3,2,1]$ or $l=[4,2,3,1]$.

Mean Position Based Rank Aggregation

As shown in the results, the Shimura technique gives worse results than the Borda's method. In an effort to improve upon the Shimura technique, we find out the mean position of each document averaged over the results of all the search engines. When sorted in an ascending order of the mean positions, a performance equivalent to that of Borda's method is obtained. In fact, it is shown that the descending order sort on the Borda score is equivalent to the ascending order sort on the mean position of the documents.

Lemma 1: The descending order sort on the Borda score is equivalent to the ascending order sort on the mean position of the documents.

Proof: Suppose a document j appears at the $i_1, i_2,..., i_N{}^{th}$ locations in the results of N search engines. Let there be M documents in all, i.e. $M=|U|$.

Then, Borda score; $B_j = (M-i_1)+(M-i_2)+...+(M-i_N) = N.M - \sum_{k=1}^{N} i_k = C-X$, where $C=N.M$ is a constant and let $X = \sum_{k=1}^{N} i_k$.

Also, Mean Score; $M_j = \dfrac{i_1+i_2+...+i_N}{N} = \dfrac{X}{N}$.

We can plot B_j and M_j against X as follows.

33

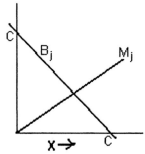

From this plot, it is easily seen that the descending order sort on the Borda score B_j is equivalent to the ascending order sort on the Mean score M_j. Hence proved.

Technique of Dubois and Prade (DP) for fuzzy ordering

As the consideration of the mean position alone gives a performance as good as that of the Borda's method, this motivates us to take the variance of the document positions also into account. With the mean $(\bar{x}_{d_i})$ and variance $(\bar{\sigma}_{d_i}^2)$ of the position of document d_i known, a Gaussian membership function is obtained as

$$\mu_{d_i}(x) = \frac{1}{\sqrt{2\pi\sigma_{d_i}^2}}\exp\left(-\frac{1}{2}\left[\frac{(x-\bar{x}_{d_i})^2}{\sigma_{d_i}^2}\right]\right) \tag{1}$$

It may be noted that the member ship function obtained from Eq. (1) would be a subnormal one. With this membership function, the technique of Dubois and Prade (DP) (Ross 1997) is adopted for aggregating the ranks. The truth value of the assertion that the document d_i occurs before d_j is given as

$$T\left(d_i \geq d_j\right) = \max_{\forall x \geq y}\left\{\min\left(\mu_{d_i}(x), \mu_{d_j}(y)\right)\right\} \tag{2}$$

Extending this to M documents, we get

$$T\left(d_i \geq \left\{d_1, d_2, ..., d_{i-1}, d_{i+1}, ..., d_M\right\}\right) = T\left(d_i \geq d_1\right) \text{ and}$$
$$T\left(d_i \geq d_2\right) \text{ and...and } T\left(d_i \geq d_{i-1}\right) \text{ and} \tag{3}$$
$$T\left(d_i \geq d_{i+1}\right) \text{ and...and } T\left(d_i \geq d_M\right).$$

where the function *and* may simply be evaluated by carrying out the minimum operation.

Example 6. Given $l_1=[3,4,2,1]$, $l_2=[2,4,3,1]$ and $l_3=[4,2,1,3]$.
Mean position of documents are

$$\bar{x}_1 = \frac{4+4+3}{3} = 3.67, \ \bar{x}_2 = 2.0, \ \bar{x}_3 = 2.67, \ \bar{x}_4 = 1.67,$$

and the variances are

$$\bar{\sigma}_1^2 = \frac{(4-3.67)^2 + (4-3.67)^2 + (3-3.67)^2}{3} = 0.22,$$

$$\bar{\sigma}_2^2 = 0.67, \ \bar{\sigma}_3^2 = 1.56, \ \bar{\sigma}_4^2 = 0.22.$$

Now using Eq. (1), we have

$$\mu_1(1) = \frac{1}{\sqrt{2\pi} \times 0.22} \exp\left(-\frac{1}{2}\left[\frac{(1-3.67)^2}{0.22}\right]\right) = 7.82 \times 10^{-8}.$$

Similarly, $\mu_1(2) = 1.5 \times 10^{-3}$, $\mu_1(3) = 0.31$, $\mu_1(4) = 0.664$,

$$\mu_2(1) = 0.231, \ \mu_2(2) = 0.48, \ \mu_2(3) = 0.231, \ \mu_2(4) = 0.025,$$

$$\mu_3(1) = 0.13, \ \mu_3(2) = 0.277, \ \mu_3(3) = 0.31, \ \mu_3(4) = 0.18,$$

$$\mu_4(1) = 0.307, \ \mu_4(2) = 0.664, \ \mu_4(3) = 0.015, \ \mu_4(4) = 4.26 \times 10^{-3}.$$

Eq. (2) now gives

$$
\begin{aligned}
T(1 \geq 2) = \max\{&\min(7.82 \times 10^{-8}, 0.231), \min(1.5 \times 10^{-3}, 0.231), \\
&\min(1.5 \times 10^{-3}, 0.48), \min(0.31, 0.231), \min(0.31, 0.48), \\
&\min(0.31, 0.231), \min(0.664, 0.231), \min(0.664, 0.48), \\
&\min(0.664, 0.231), \min(0.664, 0.025)\} = 0.48.
\end{aligned}
$$

Similarly, $T(1 \geq 3) = 0.31$, $T(1 \geq 4) = 0.664$,

$$T(2 \geq 1) = 0.231, \ T(2 \geq 3) = 0.277, \ T(2 \geq 4) = 0.48,$$

$$T(3 \geq 1) = 0.31, \ T(3 \geq 2) = 0.31, \ T(3 \geq 4) = 0.31,$$

$$T(4 \geq 1) = 0.015, \ T(4 \geq 2) = 0.48, \ T(4 \geq 3) = 0.277.$$

Finally, with Eq. (3), we get

$$T(1 \geq \{2,3,4\}) = (0.48)and(0.31)and(0.664) = 0.31,$$

$$T(2 \geq \{1,3,4\}) = 0.231, \ T(2 \geq \{1,3,4\}) = 0.31 \text{ and } T(2 \geq \{1,3,4\}) = 0.015.$$

Hence the aggregated ranking would come out to be either $l=[4,2,1,3]$ or $l=[4,2,3,1]$.

It is observed, however, that DP not only takes too long to compute but also gives poor results. An improvisation is, therefore, carried out as follows.

Membership Function Ordering (MFO) Technique

With the membership function of each document obtained as a function of positions, the amplitude of the membership function of each document at each position is evaluated. The document having the highest membership value at a given position is assigned to that position. This way, the documents are arranged in the aggregated ranking. We name this technique as membership function ordering (MFO).

From Fig. 1, we can see that out of the three documents whose subnormal membership functions are sketched, the first position has the maximum amplitude of document 1, and so document 1 must occur ahead of the rest two. Out of the remaining documents 2 and 3, the second position has the maximum amplitude of document 2, and so document 2 must be preferred over document 3 to give the ranking as document 1 $\succ$ document 2 $\succ$ document 3.

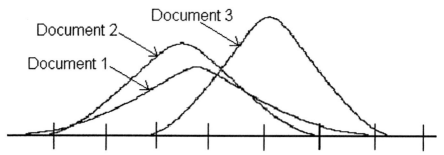

Fig. 1. Understanding MFO Technique of Rank Aggregation

Example 7. Continuing from example 6, it is seen that $\max_{i=1}^{4}\left(\mu_i(1)\right)=\mu_4(1)$.

So, for the aggregated rank l, $l[1]=4$. Similarly, $\max_{i=1}^{4}\left(\mu_i(2)\right)=\mu_4(2)$. So, $l[2]$ is also coming out to be 4. But since the document 4 has already been assigned to the first position ($l[1]=4$), we shall look for the second maximum. This way we would get $l[2]=2$. Continuing this way, the aggregated ranking is obtained as either $l=[4,2,1,3]$ or $l=[4,2,3,1]$.

Mean-By-Variance (MBV) Technique

It may be noted that if the variance of the position of two documents is same, then the document having a lesser mean position must be ranked first (see Fig. 2a). Conversely, if the mean position of two documents is same, then the document having a larger variance of position must be ranked first (see Fig. 2b). With this intuition, the Mean-By-Variance (MBV) heuristic is proposed for rank aggrega-

tion. First, the ratio $mbv(i) = \left(\dfrac{\overline{x}_{d_i}}{\overline{\sigma}_{d_i}^{\,2}}\right)$ is found for all the M documents. An as-

cending sort on the set of these fractions would give the aggregated list l.

Example 8. Continuing from example 6, $mbv(1) = \left(\dfrac{\overline{x}_1}{\overline{\sigma}_1^{\,2}}\right) =$

$\dfrac{3.67}{0.22}=16.68$, $mbv(2)=2.985$, $mbv(3)=1.71$, $mbv(4)=7.59$. On sorting these in an ascending order, we get $mbv(3) < mbv(2) < mbv(4) < mbv(1)$. Hence, the aggregated list is $l=[3,2,4,1]$.

As shown in the results, this MBV heuristic performs better than all the techniques discussed so far. However, it performs slower than MFO.

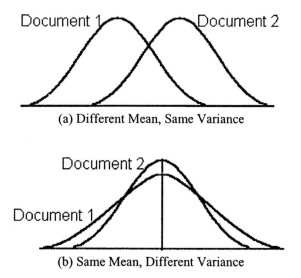

(a) Different Mean, Same Variance

(b) Same Mean, Different Variance

Fig. 2. Understanding MBV Technique of Rank Aggregation

Modified Shimura technique

As shown in the results, the Shimura technique gives worse results than the Borda's method. We, therefore, suggest an improvement in the Shimura technique, so as to enhance its performance when applied to the task of rank aggregation on the web.

We feel that the poor performance coming from the Shimura technique is primarily due to the employment of "min" function in finding $C_i = \min_{j=1}^{n} f(x_i|x_j)$, as explained earlier. The "min" function results in many ties, when a descending order sort is applied on C_i. There is no method suggested by Shimura to resolve these ties. So when resolved arbitrarily, these ties result in deterioration of the aggregated result. We, therefore, propose to replace this "min" function by an OWA operator.

Ordered Weighted Averaging (OWA) operators

The Ordered Weighted Averaging (OWA) operators were originally introduced in (Yager 1988) to provide a means of aggregation, which unifies in one operator the conjunctive and disjunctive behavior. The OWA operators, in fact, provide a parameterized family of aggregation operators including many of the well-known operators like maximum, minimum, k-order statistics, median and arithmetic mean. For some n different scores as x_1, x_2,...,x_n, the aggregation of these scores may be done using the OWA operator as follows.

$$OWA(x_1, x_2, ..., x_n) = \sum_{i=1}^{n} w_i y_i$$

where y_i is the ith largest score from amongst x_1, x_2,...,x_n. The weights are all non-negative $(\forall i, w_i \geq 0)$, and their sum equals one $\left(\sum_{i=1}^{n} w_i = 1\right)$. We note that the arithmetic mean function may be obtained using the OWA operator, if $\forall i, w_i = \dfrac{1}{n}$. Similarly, the OWA operator would yield the minimum function with $w_1=1$ and $w_i = 0$ if $i \neq 1$. The maximum function may be obtained from the OWA operator when $w_n=1$ and $w_i = 0$ if $i \neq n$.

In fact, it has been shown in (Yager 1988) that the aggregation done by the OWA operator is always between the maximum and the minimum. However, it remains to be seen what procedure should be adopted to find the values of the weights w_i. For this, we need to make use of the linguistic quantifiers, explained as follows.

Relative Fuzzy Linguistic Quantifier

A relative quantifier, $Q : [0,1] \rightarrow [0,1]$, satisfies:
$$Q(0) = 0,$$
$$\exists r \in [0,1] \text{ such that } Q(r) = 1.$$
In addition, it is non-decreasing if it has the following property:
$$\forall a,b \in [0,1], \text{ if } a > b, \text{ then } Q(a) \geq Q(b).$$
The membership function of a relative quantifier can be represented as:

$$Q(r) = \begin{cases} 0 & \text{if } r < a \\ \dfrac{r-a}{b-a} & \text{if } b \leq r \leq a \\ 1 & \text{if } r > b \end{cases},$$

where $a,b,r \in [0,1]$.

Some examples of relative quantifiers are shown in Fig. 3, where the parameters are (0.3,0.8), (0,0.5) and (0.5,1), respectively.

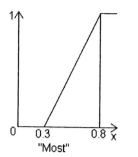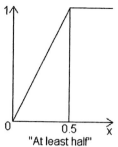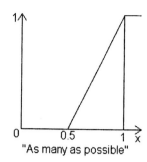

| "Most" | "At least half" | "As many as possible" |

Fig. 3. Relative Quantifiers

Yager (Yager 1988) computes the weights w_i of the OWA aggregation from the function Q describing the quantifier. In the case of a relative quantifier, with m criteria,

$$w_i = Q(i/m) - Q((i-1)/m), \ i = 1, 2, \ldots, m, \text{ with } Q(0) = 0.$$

Example 9. For the number of criteria $(m) = 7$, the fuzzy quantifier "most", with the pair $(a = 0.3, b = 0.8)$, the corresponding OWA operator would have the weighting vector as $w_1 = 0$, $w_2 = 0$, $w_3 = 0.257143$, $w_4 = 0.285714$, $w_5 = 0.285714$, $w_6 = 0.171429$, $w_7 = 0$.

We are now ready to explain the procedure of improving the Shimura technique for the problem of rank aggregation. As stated earlier, if a page spams fewer than half the participating search engines, the spam page would occupy the bottom partition of that aggregated ranking which satisfies *Condorcet* criterion. It is for this reason that we have decided to use the relative fuzzy linguistic quantifier "at least half" with the pair $(a = 0.0, b = 0.5)$ for the purpose of finding the vector C_i as follows.

$$C_i = \sum_j w_j.z_j \ ,$$

where z_j is the jth largest element in the ith row of the matrix $f(x_i|x_j)$.

Example 10. Continuing from example 5, for the number of criteria $(m) = 4$, the fuzzy quantifier "at least half" with the pair $(a = 0.0, b = 0.5)$, the corresponding OWA would have the weighting vector as $w_1 = 0.5$, $w_2 = 0.5$, $w_3 = 0.0$, $w_4 = 0.0$. This would give $C_1 = (0.5 \times 1) + (0.5 \times 0.5) + (0.0 \times 0.0) + (0.0 \times 0.0) = 0.75$, $C_2 = 1.0$, $C_3 = 1.0$, $C_4 = 1.0$.

Now, as with the Shimura technique, if a descending sort on C_i $(i=1$ to $m)$ is carried out, the sequence of $i's$ thus obtained would constitute the aggregated rank.

Entropy Minimization Technique

We go by the entropy minimization principle, which is quite useful for complex systems where the data are abundant and static (Ross 1997). We assume that we are seeking a threshold value of position for a given set of documents in the range between x_1 and x_2. We write an entropy equation for the region $p=[x_1,x]$ and

$q=[x,x_2]$. While moving an imaginary threshold value x between x_1 and x_2, the entropy is calculated for each value of x. The value of x that holds the minimum entropy is taken to be the appropriate position of the document. The entropy of document j with each value of x in the region x_1 and x_2 is expressed as follows (Ross 1997).

$$S_j(x) = p_j(x)Sp_j(x) + q_j(x)Sq_j(x)$$

where $Sp_j(x) = -\lfloor pm_j(x)\ln pm_j(x) + pr_j(x)\ln pr_j(x) \rfloor$

and $Sq_j(x) = -\lfloor qm_j(x)\ln qm_j(x) + qr_j(x)\ln qr_j(x) \rfloor$.

Here, $pm_j(x) = \dfrac{rm_j(x)+1}{r_j(x)+1}$,

$$pr_j(x) = \frac{rr_j(x)+1}{r_j(x)+1},$$

$$qm_j(x) = \frac{Rm_j(x)+1}{R_j(x)+1},$$

$$qr_j(x) = \frac{Rr_j(x)+1}{R_j(x)+1},$$

$$R_j(x) = Rm_j(x) + Rr_j(x),$$

$$r_j(x) = rm_j(x) + rr_j(x)$$

$$p_j(x) = \frac{r_j(x)}{r},$$

$$q_j(x) = 1 - p_j(x),$$

$Rm_j(x) =$ Number of samples of document j in $[x_1+x,x_2]$,

$Rr_j(x) =$ Number of samples of documents other than j in $[x_1+x,x_2]$,

$rm_j(x) =$ Number of samples of document j in $[x_1,x_1+x]$,

$rr_j(x) =$ Number of samples of documents other than j in $[x_1,x_1+x]$,

and $r =$ Total number of samples in $[x_1,x_2]$.

The value of x that holds the minimum value of entropy $S_j(x)$ would be taken as the position of the document j.

Example 11. Given $l_1=[3,4,2,1]$, $l_2=[2,4,3,1]$ and $l_3=[4,2,1,3]$, we see that $r = 12$.

For document 1, $Rm_1(1)=3$, $Rr_1=9$, $R_1=3+9=12$,

$rm_1(1)=0$, $rr_1(1)=0$, $r_1(1)=0+0=0$,

$$pm_1(1) = \frac{0+1}{0+1} = 1, \quad pr_1(1) = \frac{0+1}{0+1} = 1,$$

$$qm_1(1) = \frac{3+1}{12+1} = 0.3077, \quad qr_1(1) = \frac{9+1}{12+1} = 0.769,$$

$$p_1(1) = \frac{0}{12} = 0, \quad q_1(1)=1-0=1,$$

$Sp_1(1)=-[1\times\ln(1)+1\times\ln(1)]=0$,

$Sq_1(1)=-[0.3077\times\ln(0.3077)+0.769\times\ln(0.769)]=0.5645$,

$S_1(1)=0\times0+1\times0.5645=0.5645$.

Similarly, $S_1(2)=0.5488$, $S_1(3)=0.4588$, $S_1(4)=0.4531$.

Since the entropy of the first document is minimum at the 4th position, the document 1 must be placed in the last position of the aggregated rank l.

On the same lines, for document 2, it is found that $S_2(1)=0.5645$, $S_2(2)=0.5454$, $S_2(3)=0.5468$.

For document 3, $S_3(1)=0.5645$, $S_3(2)=0.5454$, $S_3(3)=0.5468$, and for document 4, $S_4(1)=0.5645$, $S_4(2)=0.5454$, $S_4(3)=0.4588$, $S_4(4)=0.5488$.

It may be seen that the document 4 must occupy the 3rd position, while documents 2 and 3 are tied for the 2nd position.

Hence, the aggregated rank is either $l=[3,2,4,1]$ or $l=[2,3,4,1]$.

Experiments and Results

For experimentation, we have taken the same 37 queries as in (Dwork et al 2001), which in turn, are the superset of the queries used in (Bharat and Henzinger 1998) and (Chakrabarti et al 1998). These queries are:

affirmative action, alcoholism, amusement parks, citrus groves, classical guitar, architecture, bicycling, blues, cheese, computer vision, cruises, Death Valley, field hockey, gardening, graphic design, Gulf war, HIV, java, Lipari, lyme disease, mutual funds, National parks, parallel architecture, Penelope Fitzgerald, recycling cans, rock climbing, San Francisco, Shakespeare, stamp collecting, sushi, table tennis, telecommuting, Thailand tourism, vintage cars, volcano, zen buddhism, and *Zener.*

We are also using the same seven search engines as in (Dwork et al 2001), viz. *Altavista* (AltaVista 2001), *Alltheweb* (AlltheWeb 2001), *Excite* (Excite 2001), *Google* (Google 2001), *Hotbot* (HotBot 2001), *Lycos* (Lycos 2001) and *Northernlight* (Northernlight 2001). Falling in line with (Dwork et al 2001), we are taking only the top 100 results from these search engines, aggregating these results using the techniques discussed in this chapter and then comparing the aggregated list with the one obtained using Borda's method.

The results are summarized in Table 1. For each of the seven techniques of rank aggregation discussed in this paper, we give the spearman footrule distance between the aggregated list thus obtained and the result listings of the seven search engines, averaged for the 37 test queries. An estimate of the run time of each of these techniques of rank aggregation is also given relative to the Borda's method.

Table 1. Performance Comparison of Different Rank Aggregation Techniques Averaged over the Given Set of 37 Test Queries

Aggregation Technique	Aggregated Distance [0,1]	Time Taken Relative to Borda
Borda	0.513214	1.0
Shimura	0.529648	3.8
DP	0.571651	30640.07
MFO	0.513111	228.73
MBV	0.512847	917.83
Improved Shimura	0.512028	86.11
Entropy	0.510528	5.33

From Table 1, it may be seen that the least distance is obtained in the case of Entropy based technique and hence the best performance. That is followed by Improved Shimura, MBV, MFO, Borda, Shimura and DP, in that order. In terms of run time taken, Borda is still the best, followed by Shimura, Entropy, Improved Shimura, MFO, MBV and DP, in that order. It is quite clear that DP is by far the worst performer, both in terms of the aggregated distance as well as the run time, and hence should never be used. The overall best performance appears to be coming from the Entropy-based technique, followed by Improved Shimura. MFO and MBV are almost a joint-third. The classical Borda's method is better than the Shimura technique in both respects. From Table 2 through Table 8, we can see how relevant the results get after applying the different rank aggregation techniques discussed in this paper. These results were taken in September 2001.

Table 2. Top Few Results for the Query *parallel architecture* after Rank Aggregation using the Borda's Method

Sequence	URL
1.	www.npac.syr.edu/
2.	www.cis.ohio-state.edu/~panda/pac.html
3.	www.cc.gatech.edu/computing/Architecture/arch.html
4.	atanasoff.nmsu.edu/
5.	pertsserver.cs.uiuc.edu/iclass/r1995/node3.html
6.	www.eecg.toronto.edu/~tcm/CourseECE1755.html
7.	tracebase.nmsu.edu/
8.	www.c3.lanl.gov/cic19/teams/par_arch/par_arch.shtml
9.	classic.korea.ac.kr
10.	www-didc.lbl.gov/DPSS/s/code/

Table 3. Top Few Results for the Query *parallel architecture* after Rank Aggregation using the Shimura Technique

Sequence	URL
1.	www.npac.syr.edu/
2.	www.cis.ohio-state.edu/~panda/pac.html
3.	classic.korea.ac.kr/
4.	www.pdos.lcs.mit.edu/
5.	www.cs.cmu.edu/~scandal/research-groups.html
6.	www.jjg.net/ia/
7.	www.cs.rice.edu/~roth/conferences.html
8.	www.nersc.gov/research/FTG/via/
9.	www.etl.go.jp/~6822/index.html
10.	www.phy.duke.edu/brahma/

Table 4. Top Few Results for the Query *parallel architecture* after Rank Aggregation using the DP Technique

Sequence	URL
1.	cardit.et.tudelft.nl/~heco/chess/p/p.html
2.	www.ai.sri.com/%7eoaa/main.html
3.	www.cs.purdue.edu/research/PaCS/parasol.html
4.	presto.stsci.edu/stsci/meetings/irw/proceedings/poovendranr.dir/poovendranr.html
5.	www.imv.es/marcos/html/ficha000.htm
6.	pweb.sophia.ac.jp/~y-tatsuz/rinkou/index.htm
7.	www.ime.usp.br/~enec/eventos/icpact98.html
8.	www.aps.org/BAPSDPP98/abs/S3100014.html
9.	acoma.santafe.edu/sfi/publications/Abstracts/92-11-055abs.html
10.	citforum.gatchina.ru/hardware/articles/art_5.shtml

Table 5. Top Few Results for the Query *parallel architecture* after Rank Aggregation using the MFO Technique

Sequence	URL
1.	www.npac.syr.edu/
2.	www.cis.ohio-state.edu/~panda/pac.html
3.	www.cc.gatech.edu/computing/Architecture/arch.html
4.	atanasoff.nmsu.edu/
5.	pertsserver.cs.uiuc.edu/iclass/r1995/node3.html
6.	www.eecg.toronto.edu/~tcm/CourseECE1755.html
7.	tracebase.nmsu.edu/
8.	www.c3.lanl.gov/cic19/teams/par_arch/par_arch.shtml

9.	classic.korea.ac.kr
10.	www-didc.lbl.gov/DPSS/s/code/

Table 6. Top Few Results for the Query *parallel architecture* after Rank Aggregation using the MBV Technique

Sequence	URL
1.	www.npac.syr.edu/
2.	www.cis.ohio-state.edu/~panda/pac.html
3.	www.cc.gatech.edu/computing/Architecture/arch.html
4.	atanasoff.nmsu.edu/
5.	pertsserver.cs.uiuc.edu/iclass/r1995/node3.html
6.	www.eecg.toronto.edu/~tcm/CourseECE1755.html
7.	tracebase.nmsu.edu/
8.	www.c3.lanl.gov/cic19/teams/par_arch/par_arch.shtml
9.	classic.korea.ac.kr
10.	www-didc.lbl.gov/DPSS/s/code/

Table 7. Top Few Results for the Query *parallel architecture* after Rank Aggregation using the Improved Shimura Method

Sequence	URL
1.	www.npac.syr.edu/
2.	www.cis.ohio-state.edu/~panda/pac.html
3.	www.cc.gatech.edu/computing/Architecture/arch.html
4.	atanasoff.nmsu.edu/
5.	pertsserver.cs.uiuc.edu/iclass/r1995/node3.html
6.	www.eecg.toronto.edu/~tcm/CourseECE1755.html
7.	tracebase.nmsu.edu/
8.	www.c3.lanl.gov/cic19/teams/par_arch/par_arch.shtml
9.	classic.korea.ac.kr
10.	www-didc.lbl.gov/DPSS/s/code/

Table 8. Top Few Results for the Query *parallel architecture* after Rank Aggregation using the Entropy Minimization Technique

Sequence	URL
1.	pcl.cs.ucla.edu/projects/parsec/manual/
2.	ads.computer.org/concurrency/pd1996/p4010abs.htm
3.	www.jjg.net/ia/
4.	www.cs.rice.edu/~roth/conferences.html
5.	www.etl.go.jp/~6822/index.html
6.	www.phy.duke.edu/brahma/
7.	wotug.ukc.ac.uk/parallel/

8.	linas.org/linux/i370.html
9.	williamstallings.com/COA5e.html
10.	ei.cs.vt.edu/~history/Parallel.html

From Table 2 through Table 8, the readers are invited to see for themselves that the results obtained from the entropy based rank aggregation technique are the most relevant ones. It may, however, appear that the results obtained from MBV, MFO and Improved Shimura are identical to those obtained from the Borda's method. But then, it is just one of the 37 queries used in our experiments, and that too the top 10 results of a total of 591. For other queries, the results of MBV, MFO and Improved Shimura are found to be better than that of the Borda's method. As is already seen from Table 1, the average performance of MBV, MFO and Improved Shimura is better than the Borda's method.

Summary

We have employed fuzzy techniques for finding an aggregated rank having a minimum footrule distance with the given lists of documents obtained from different search engines, in response to a given query. First, the performance of classical fuzzy ranking techniques like that of Shimura's and the other one due to Dubois and Prade (DP) were adopted for the problem of rank aggregation, but both of them were found to be rather hopeless. However, taking lessons out of their comparison studies, a few new heuristics, namely, MFO, MBV, Improved Shimura and Entropy-based are then proposed in this chapter. MFO works by comparing the values of the membership functions of the document positions. MBV proceeds by carrying out an ascending sort on the ratio of mean and variance of the document positions. The Shimura technique is improved by replacing the *min* function with the OWA operation. The entropy-based technique goes about by adopting the entropy minimization principle for the purpose of rank aggregation.

Our experimental procedure falls in line with the ones found in literature. Borda's method is found to be better than both Shimura as well as DP. DP is clearly the worst performer, followed by Shimura. MBV renders a lesser footrule distance but takes more time than MFO. The Improved Shimura technique not only gives lesser distance but also takes lesser time as compared to both MBV and MFO. However, the entropy-based technique emerges out to be the best performer in time and distance, when compared to MBV, MFO as well as Improved Shimura.

References

Ahmad N, Beg MMS (2002) Fuzzy Logic Based Rank Aggregation Methods for the World Wide Web. In: Proc. International Conference on Artificial Intelligence in Engineering and Technology (ICAIET 2002), Kota Kinabalu, Malaysia, June 17-18, 2002.

AlltheWeb(2001) URL http://www.alltheweb.com. September 2001.

AltaVista (2001) URL http://www.altavista.com. September 2001.

Beg MMS, Ahmad N (2002) Genetic Algorithm Based Rank Aggregation for the Web. In: Proc. 6th International Conference on Computer Science and Informatics - a track at the 6th Joint Conference on Information Sciences (JCIS 2002), March 8-13, 2002, Durham, NC, USA, pp 329-333.

Bharat K, Henzinger M (1998) Improved Algorithms for Topic Distillation in a Hyperlinked Environment. In: Proceedings of ACM SIGIR, pp 104-111.

Borda JC (1781) Memoire sur les election au scrutin. Histoire de l'Academie Royale des Sciences.

Chakrabarti S, Dom B, Gibson D, Kumar R, Raghavan P, Rajagopalan S, Tomkins A (1998) Experiments in topic distillation. In: Proceedings of ACM SIGIR Workshop on Hypertext Information Retrieval on the Web.

Diaconis P, Graham R (1977) Spearman's Footrule as a Measure of Disarray. Journal of the Royal Statistical Society Series B, 39(2):262-268.

Dwork C, Kumar R, Naor M, Sivakumar D (2001) Rank Aggregation Methods for the Web. In: Proceedings of the Tenth World Wide Web Conference. Hong Kong.

Excite (2001) URL http://www.excite.com. September 2001.

Google (2001) URL http://www.google.com. September 2001.

HotBot (2001) URL http://www.hotbot.com. September 2001.

Lycos (2001) URL http://www.lycos.com. September 2001.

NorthernLight (2001) URL http://www.northernlight.com. September 2001.

Ross TJ (1997) Fuzzy Logic with Engineering Applications. Tata McGraw Hill.

Shimura M (1973) Fuzzy Sets Concept in Rank-Ordering Objects. J. Math. Anal. Appl. 43:717-733.

Yager RR (1988) On Ordered Weighted Averaging Aggregation Operators in Multicriteria Decision Making. IEEE Trans. Systems, Man and Cybernetics, vol. 18, no. 1, January/February, pp 183-190.

Automatic Learning of Multiple Extended Boolean Queries by Multiobjective GA-P Algorithms*

O. Cordón[1], F. Moya[2], and C. Zarco[3]

[1] Dept. of Computer Science and Artificial Intelligence
University of Granada. 18071 - Granada (Spain)
`ocordon@decsai.ugr.es`
[2] Dept. of Information Sciences. Faculty of Information Sciences
University of Granada. 18071 - Granada (Spain)
`felix@ugr.es`
[3] PULEVA Salud S.A.
Camino de Purchil, 66. 18004 - Granada (Spain)
`czarco@puleva.es`

Summary. In this contribution, a new Inductive Query by Example process is proposed to automatically derive extended Boolean queries for fuzzy information retrieval systems from a set of relevant documents provided by a user. The novelty of our approach is that it is able to simultanously generate several queries with a different precision-recall tradeoff in a single run. To do so, it is based on an advanced evolutionary algorithm, GA-P, specially designed to tackle with multiobjective problems by means of a Pareto-based multiobjective technique. The performance of the new proposal will be tested on the usual Cranfield collection and compared to the well-known Kraft et al.'s process.

Key words: Fuzzy Information Retrieval, Inductive Query by Example, Evolutionary Algorithms, Genetic Algorithm-Programming, Multiobjective Evolutionary Algorithms.

1 Introduction

Information retrieval (IR) may be defined, in general, as the problem of the selection of documentary information from storage in response to search questions provided by a user [28, 2]. Information retrieval systems (IRSs) are a

---

* This work has been supported by CICYT under Project TIC2002-03276 and by the University of Granada under Project "Mejora de Metaheurísticas mediante Hibridación y sus Aplicaciones".

kind of information system that deal with data bases composed of information items – documents that may consist of textual, pictorial or vocal information – and process user queries trying to allow the user to access to relevant information in an appropriate time interval.

The underlying retrieval model of most of the commercial IRSs is the Boolean one [34], which presents some limitations. Due to this fact, some paradigms have been designed to extend this retrieval model and overcome its problems, such as the vector space [28] or the fuzzy information retrieval (FIR) models [3, 12].

However, the increase in the power of the retrieval model also comes with a high complexity augment in the query language, what makes difficult for the user to represent his information needs in the form of a valid query. This is especially significant in the case of fuzzy IRSs, whose query language allows us to formulate weighted Boolean (fuzzy) queries where the query terms are joined by the logical operators AND and OR. If it is difficult for a human user to formulate a classical Boolean query due to the need to know how to properly connect the query terms together using the Boolean operators, it will be even more difficult to both define the query structure and specify the query term weights to retrieve the desired documents.

Hence, the paradigm of Inductive Query by Example (IQBE) [5], where queries describing the information contents of a set of documents provided by a user are automatically derived, can be useful to solve this problem and assist the user in the query formulation process. Focusing on the FIR model, the most known existing approach is that of Kraft et al. [24], which is based on genetic programming [23]. Moreover, several other approaches have been proposed based on the use of more advanced evolutionary algorithms [1], such as genetic algorithm-programming (GA-P) [21] or simulated annealing-programming (SA-P) [30], in order to improve the Kraft et al.'s one [8, 9, 10].

On the other hand, it is well known that the performance of an IRS is usually measured in terms of two different criteria, precision and recall [34]. This way, the optimization of any of its components, and concretely the automatic learning of fuzzy queries, is a clear example of a multiobjective problem. Usually, the application of evolutionary algorithms in the area has been based on combining both criteria in a single scalar fitness function by means of a weighting scheme [7]. However, there is a kind of evolutionary algorithms specially designed for multiobjective problems, *multiobjective evolutionary algorithms*, which are able to obtain different non-dominated solutions to the problem in a single run [14, 6].

In [11], it was proposed an extension of Smith and Smith's IQBE algorithm to learn Boolean queries [33] transforming it into a Pareto-based multiobjective evolutionary algorithm. The proposed process obtained very good results in one of the most known IR benchmarks, the Cranfield document collection.

In this chapter, the same idea will be applied to learn extended Boolean queries for FIRSs. This way, an IQBE process similar to that proposed in [8] – based on a GA-P algorithm – will be transformed in a Pareto-based multiobjective evolutionary algorithm to deal with this problem. The proposal will be compared to the previous approach by Kraft et al. in the well-known Cranfield documentary base.

To do so, this paper is structured as follows. In Section 2, some preliminaries are introduced by reviewing the basis of IRSs and FIRSs, IQBE and multiobjective evolutionary algorithms. Then, two single-objective IQBE processes are discussed in Section 3, Kraft et al.'s genetic programming-based algorithm and our previous proposal based on the use of the more advanced GA-P technique. Section 4 is devoted to extend the latter process to deal with the multiobjective problem of simultaneously optimizing both precision and recall by means of the Pareto-based approach. The experiments developed on the Cranfield collection are presented and analyzed in Section 5. Finally, Section 6 summarizes several concluding remarks.

2 Preliminaries

2.1 Boolean Information Retrieval Systems

An IRS is basically constituted by three main components, as showed in Figure 1:

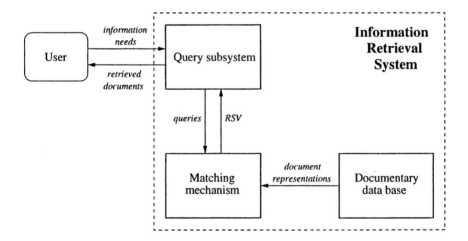

Fig. 1. Generic structure of an information retrieval system

1. A *documentary data base*, which stores the documents and the representation of their information contents. It is associated with the *indexer module*,

which automatically generates a representation for each document by extracting the document contents. Textual document representation is typically based on index terms (that can be either single terms or sequences) which are the content identifiers of the documents.

2. A *query subsystem*, which allows the users to formulate their queries and presents the relevant documents retrieved by the system to them. To do so, it includes a *query language*, that collects the rules to generate legitimate queries and procedures to select the relevant documents.

3. A *matching or evaluation mechanism*, which evaluates the degree to which the document representations satisfy the requirements expressed in the query, the so called *retrieval status value* (RSV), and retrieves those documents that are judged to be relevant to it.

In the Boolean retrieval model, the indexer module performs a binary indexing in the sense that a term in a document representation is either significant (appears at least once in it) or not (it does not appear in it at all). Let D be a set of documents and T be a set of unique and significant terms existing in them. The indexer module of the Boolean IRS defines an indexing function:

$$F : D \times T \to \{0, 1\}$$

where $F(d, t)$ takes value 1 if term t appears in document d and 0 otherwise.

On the other hand, user queries in this model are expressed using a query language that is based on these terms and considers combinations of simple user requirements with logical operators AND, OR and NOT [28, 34]. The result obtained from the processing of a query is a set of documents that totally match with the query, i.e., only two possibilities are considered for each document: to be (RSV=1) or not to be (RSV=0) relevant for the user's needs, represented by the user query.

Thus, the Boolean model presents several problems that correspond to the different Boolean IRS components such as:

- It does not provide the user with tools to express the degree of relevance of the index terms to the documents (*indexer module*).
- It has no method to express a user's judgement of the importance of the terms in the query (*query language*).
- There are no partial degrees of relevance of documents to queries possibly useful in ranking (*matching mechanism*).

2.2 Fuzzy Information Retrieval Systems

FIRSs make use of the fuzzy set theory [35] to deal with the imprecision and vagueness that characterizes the IR process. As stated in [3], the use of fuzzy sets in IR is suitable due to two main reasons:

1. It is a formal tool designed to deal with imprecision and vagueness.

2. It facilitates the definition of a superstructure of the Boolean model, so that existing Boolean IRSs can be modified without redesigning them completely.

Hence, trying to solve the previously introduced problems of the Boolean IR model, FIR mainly extends it in three aspects:

1. Document representations become fuzzy sets defined in the universe of terms, and terms become fuzzy sets defined in the universe of discourse of documents, thus introducing a degree of relevance (aboutness) between a document and a term.
2. Numeric weights (and in recent proposals, linguistic terms [3, 20]) are considered in the query with different semantics (a review of them is to be found in [3]), thus allowing the user to quantify the "subjective importance" of the selection requirements.
3. Since the evaluation of the relevance of a document to a query is also an imprecise process, a degree of document relevance is introduced, i.e., the RSV is defined as a real value in [0,1]. To do so, the classical complete matching approach and Boolean set operators are modeled by means of fuzzy operators appropriately performing the matching of queries to documents in a way that preserves the semantics of the former.

Thus, the operation mode of the three components of an FIRS is showed as follows.

Indexer Module

The indexer module of the FIRS defines an indexing function which maps the document-term pair into the real interval [0,1]:

$$F : D \times T \rightarrow [0, 1]$$

It can be seen that F is the membership function of a two-dimensional fuzzy set (a fuzzy relation) mapping the degree to which document d belongs to the set of documents "about" the concept(s) represented by term t. By projecting it, a fuzzy set can be associated to each document and term:

$$d_i = \{< t, \mu_{d_i}(t) > | t \in T\} \quad ; \quad \mu_{d_i}(t) = F(d_i, t)$$

$$t_j = \{< d, \mu_{t_j}(d) > | d \in D\} \quad ; \quad \mu_{t_j}(d) = F(d, t_j)$$

There are different ways to define the indexing function F. In this paper, we will work with the normalized *inverse document frequency* [28]:

$$w_{d,t} = f_{d,t} \cdot log(N/N_t) \quad ; \quad F(d,t) = \frac{w_{d,t}}{Max_d \, w_{d,t}}$$

where $f_{d,t}$ is the frequency of term t in document d, N is the number of documents in the collection and N_t is the number of documents where term t appears at least once.

Matching mechanism

It operates in a different way depending on the interpretation associated to the numeric weights included in the query (the interested reader can refer to [3, 12] to get knowledge about the three existing approaches). In this paper, we consider the *importance* interpretation, where the weights represent the relative importance of each term in the query.

In this case, the RSV of each document to a fuzzy query q is computed as follows [29]. When a single term query is logically connected to another by means of the AND or OR operators, the relative importance of the single term in the compound query is taken into account by associating a weight to it. To maintain the semantics of the query, this weighting has to take a different form according as the single term queries are ANDed or ORed. Therefore, assuming that A is a fuzzy term with assigned weight w, the following expressions are applied to obtain the fuzzy set associated to the weighted single term queries A_w (in the case of *disjunctive queries*) and A^w (for *conjunctive ones*):

$$A_w = \{< d, \mu_{A_w}(d) > \,|\, d \in D\} \qquad ; \qquad \mu_{A_w}(d) = Min\,(w, \mu_A(d))$$

$$A^w = \{< d, \mu_{A^w}(d) > \,|\, d \in D\} \qquad ; \qquad \mu_{A^w}(d) = Max\,(1 - w, \mu_A(d))$$

On the other hand, if the term is negated in the query, a negation function is applied to obtain the corresponding fuzzy set:

$$\overline{A} = \{< d, \mu_{\overline{A}}(d) > \,|\, d \in D\} \quad ; \quad \mu_{\overline{A}}(d) = 1 - \mu_A(d)$$

Once all the single weighted terms involved in the query have been evaluated, the fuzzy set representing the RSV of the compound query is obtained by combining the partial evaluations into a single fuzzy set by means of the following operators:

$$A\ AND\ B = \{< d, \mu_{A\ AND\ B}(d) > \,|\, d \in D\}$$
$$\mu_{A\ AND\ B}(d) = Min(\mu_A(d), \mu_B(d))$$

$$A\ OR\ B = \{< d, \mu_{A\ OR\ B}(d) > \,|\, d \in D\}$$
$$\mu_{A\ OR\ B}(d) = Max(\mu_A(d), \mu_B(d))$$

We should note that all the previous expressions can be generalized to work with any other t-norm, t-conorm and negation function different from the usual minimum, maximum and one-minus function. In this contribution, we will consider the former ones.

Query Subsystem

It affords a fuzzy set q defined on the document domain specifying the degree of relevance of each document in the data base with respect to the processed query:

$$q = \{< d, \mu_q(d) > \,|\, d \in D\} \quad ; \quad \mu_q(d) = RSV_q(d)$$

Thus, one of the advantages of FIRSs is that documents can be ranked in order to the membership degrees of relevance – as in IRSs based on the vector space model [28] – before being presented to the user as query response. The final relevant document set can be specified by him in two different ways: providing an upper bound for the number of retrieved documents or defining a threshold σ for the relevance degree (as can be seen, the latter involves obtaining the σ-cut of the query response fuzzy set q).

Focusing on the latter approach, which will be the one considered in this paper, the final set of documents retrieved would be:

$$R = \{d \in D \,|\, RSV_q(d) \geq \sigma\}$$

2.3 Inductive Query by Example

IQBE was proposed in [5] as "a process in which searchers provide sample documents (examples) and the algorithms induce (or learn) the key concepts in order to find other relevant documents". This way, IQBE is a process for assisting the users in the query formulation process performed by machine learning methods [26]. It works by taking a set of relevant (and optionally, non relevant documents) provided by a user – that can be obtained from a preliminary query or from a browsing process in the documentary base – and applying an off-line learning process to automatically generate a query describing the user's needs (as represented by the document set provided by him). The obtained query can then be run in other IRSs to obtain more relevant documents. This way, there is no need that the user interacts with the process as in other query refinement techniques such as relevance feedback [28, 2].

Several IQBE algorithms have been proposed for the different existing IR models. On the one hand, Smith and Smith [33] introduced a Boolean query learning process based on genetic programming. Besides, a similar idea to that proposed in this paper was applied in [11] in order to allow the Smith and Smith's algorithm to simultaneously derive multiple Boolean queries from the same document set. On the other hand, all of the machine learning methods considered in Chen et al.'s paper [5] (regression trees, genetic algorithms and simulated annealing) dealt with the vector space model. Moreover, there are several approaches for the derivation of weighted Boolean queries for FIRSs, such as the genetic programming algorithm of Kraft et al. [24], that will be reviewed in the next section, our niching GA-P method [9] and our SA-P method [10], based on a simulated annealing-genetic programming hybrid.

For descriptions of some of the previous techniques based on EAs refer to [8, 10].

2.4 Multiobjective Evolutionary Algorithms

Evolutionary computation uses computational models of evolutionary processes as key elements in the design and implementation of computer-based problem solving systems. There are a variety of evolutionary computational models that have been proposed and studied which are referred as *evolutionary algorithms* (EAs) [1]. Concretely, four well-defined EAs have served as the basis for much of the activity in the field: *genetic algorithms* (GAs) [25], *evolution strategies* [32], *genetic programming* (GP) [23] and *evolutionary programming* [16].

An EA maintains a population of trial solutions, imposes random changes to these solutions, and incorporates selection to determine which ones are going to be maintained in future generations and which will be removed from the pool of the trials. But there are also important differences between them. Focusing on the two kinds of EAs considered on this paper, GAs and GP, the former emphasize models of genetic operators as observed in nature, such as crossover (recombination) and mutation, and apply these to abstracted chromosomes with different representation schemes according to the problem being solved. As regards GP, it constitutes a variant of GAs, based on evolving structures encoding programs such as expression trees. Apart from adapting the crossover and mutation operators to deal with the specific coding scheme considered, the rest of the algorithm components remain the same.

EAs are very appropriate to solve multiobjective problems. These kinds of problems are characterized by the fact that several objectives have to be simultaneously optimized. Hence, there is not usually a single best solution solving the problem, i.e. being better than the remainder with respect to every objective, as in single-objective optimization. Instead, in a typical multiobjective optimization problem, there is a set of solutions that are superior to the remainder when all the objectives are considered, the *Pareto* set. These solutions are known as *non-dominated solutions* [4], while the remainder are known as *dominated solutions*. Since none of the Pareto set solutions is absolutely better than the other non-dominated solutions, all of them are equally acceptable as regards the satisfaction of all the objectives.

This way, thanks to the use of a population of solutions, EAs can search many Pareto-optimal solutions in the same run. Generally, multiobjective EAs only differ from the rest of EAs in the fitness function and/or in the selection mechanism. The evolutionary approaches in multiobjective optimization can be classified in three groups: *plain aggregating approaches, population-based non-Pareto approaches*, and *Pareto-based approaches* [14, 6].

The first group constitutes the extension of classical methods to EAs. The objectives are artificially combined, or aggregated, into a scalar function according to some understanding of the problem, and then the EA is applied in the usual way[4]. Optimizing a combination of the objectives has the advantage

---

[4] As said, this has been the approach usually followed in the application of EAs to IR.

of producing a single compromise solution but there are two problems: i) it can be difficult to define the combination weights in order to obtain acceptable solutions, and ii) if the optimal solution generated can not be finally accepted, new runs of the EA may be required until a suitable solution is found.

Population-based non-Pareto approaches allow to exploit the special characteristics of EAs. A non-dominated individual set is obtained instead of generating only one solution. In order to do so, the selection mechanism is changed. Generally, the best individuals according to each of the objectives are selected, and then these partial results are combined to obtain the new population. An example of a multiobjective GA of this group is Vector Evaluated Genetic Algorithm (VEGA) [31].

Finally, *Pareto-based approaches* seem to be the most active research area on multiobjective EAs nowadays. In fact, algorithms included within this family are divided in two different groups: first and second generation [6]. They all attempt to promote the generation of multiple non-dominated solutions, as the former group, but directly making use of the Pareto-optimality definition. To introduce this concept, let us consider, without loss of generality, a multiobjective minimization problem with m parameters (decision variables) and n objectives:

$$Min \ f(x) = (f_1(x), f_2(x), \ldots, f_n(x)), \ \ with \ x = (x_1, x_2, \ldots, x_m) \in X$$

A decision vector $a \in X$ dominates another $b \in X$ if, and only if:

$$\forall i \in 1, 2, \ldots, n \mid f_i(a) \le f_i(b) \ \ \land \ \ \exists j \in 1, 2, , \ldots, n \mid f_j(a) < f_j(b)$$

As said, any vector that is not dominated by any other is said to be *Pareto-optimal* or *non-dominated*.

This way, to calculate the probability of reproduction of each individual in this approach, the solutions are compared by means of the dominance relation. Different equivalence groups are defined depending on the dominance of their constituent individuals among the remainder and those individuals belonging to the "good" classes (those groups including individuals dominating a large number of the remainder) are assigned a higher selection probability than "bad" classes.

The difference between the first and the second generation of Pareto-based approaches arise on the use of elitism. Algorithms included within the first generation group, such as Niched Pareto Genetic Algorithm (NPGA), Non-dominated Sorting Genetic Algorithm (NSGA) and Multiple-Objective Genetic Algorithm (MOGA) (the one considered in this contribution), do not consider this characteristic. On the other hand, second generation Pareto-based multiobjective EAs are based on the consideration of an auxiliary population where the non-dominated solutions generated among the different iterations are stored. Examples of the latter family are Strenght Pareto EA (SPEA) and SPEA2, NSGA2 and NPGA2, among others. As can be seen, several of the latter algorithms are elitist versions of the corresponding first generation ones. For the description of all of these algorithms, the interested reader can refer to [14, 6].

Finally, it is important to notice that, although the Pareto-based ranking correctly assigns all non-dominated individuals the same fitness, it does not guarantee that the Pareto set is uniformly sampled. When multiple equivalent optima exist, finite populations tend to converge to only one of them, due to stochastic errors in the selection process. This phenomenom is known as *genetic drift* [13]. Since preservation of diversity is crucial in the field of multiobjective optimization, several multiobjective EAs have incorporated the *niche* and *species* concepts [18] for the purpose of favouring such behaviour. We will also consider this aspect in our proposal.

3 Single-objective IQBE Processes for Extended Boolean Queries

In this section, two different IQBE algorithms to learn extended Boolean queries are reviewed. First, the well known Kraft et al.'s GP-based process, which will be used for comparison purposes in this contribution, is presented in the next subsection. Then, a variant involving the use of the GA-P algorithm is analyzed in Section 3.2. This latter algorithm will the one extended in Section 4 to build the multiobjective proposal introduced in this paper.

3.1 The Kraft et al.'s Genetic Programming-based IQBE Algorithm for Fuzzy Information Retrieval Systems

In [24], Kraft et al. proposed an IQBE process to deal with extended Boolean queries in FIRSs. The algorithm is based on GP and its components are described next[5].

Coding Scheme

The fuzzy queries are encoded in expression trees, whose terminal nodes are query terms with their respective weights and whose inner nodes are the Boolean operators *AND, OR* or *NOT*.

Selection Scheme

It is based on the classical generational scheme, together with the elitist selection. The intermediate population is created from the current one by means of Tournament selection [25], which involves the random selection of a number t of individuals from the current population and the choice of the best adapted of them to take one place in the new population.

---

[5] Notice that the composition of several components is not the original one proposed by Kraft et al. but they have been changed in order to improve the algorithm performance. Of course, the basis of the process have been maintained.

Genetic Operators

The usual GP crossover is considered [23], which is based on randomly selecting one edge in each parent and exchanging both subtrees from these edges between the both parents.

On the other hand, the following three possibilities are randomly selected – with the showed probability – for the GP mutation:

a) Random selection of an edge and random generation of a new subtree that substitutes the old one located in that edge (p=0.4).
b) Random change of a query term for another one, not present in the encoded query, but belonging to any relevant document (p=0.1).
c) Random change of the weight of a query term (p=0.5).

For the latter case, *Michalewicz's non-uniform mutation operator* [25] is considered. It is based on making a uniform search in the initial space in the early generations, and a very local one in later stages. Let w be the query weight selected for mutation (the domain of w is $[0, 1]$), the new value for it is:

$$w' = \begin{cases} w + \Delta(t, 1 - w), & \text{if } a = 0 \\ w - \Delta(t, w), & \text{if } a = 1 \end{cases}$$

where $a \in \{0, 1\}$ is a random number and the function $\Delta(t, y)$ returns a value in the range $[0, y]$ such that the probability of $\Delta(t, y)$ being close to 0 increases as the number of generations increases.

Generation of the Initial Population

A first individual is obtained by generating a random tree representing a query with a maximum predefined length and composed of randomly selected terms existing in the initial relevant documents provided by the user, and with all the term weights set to 1. The remaining individuals are generated in the same way but with a random size and random weights in [0,1].

Fitness function

Two different possibilities are considered based on the classical precision and recall measures (to get more information about them, see [34]):

$$F_1 = \frac{\sum_d r_d \cdot f_d}{\sum_d r_d} \quad ; \quad F_2 = \alpha \cdot \frac{\sum_d r_d \cdot f_d}{\sum_d f_d} + \beta \cdot \frac{\sum_d r_d \cdot f_d}{\sum_d r_d}$$

with $r_d \in \{0, 1\}$ being the relevance of document d for the user and $f_d \in \{0, 1\}$ being the retrieval of document d in the processing of the current query. Hence, F_1 only considers the recall value obtained by the query, while F_2 also takes its precision into account.

Moreover, as simple queries are always prefered by the user, a selection criterion has been incorporated to the algorithm in order to consider more fitted those queries with a lesser complexity among a group of chromosomes with the same fitness value.

3.2 A GA-P-based Extension of Kraft et al.'s Method

Although the algorithm proposed by Kraft et al. analyzed in the previous section obtains good results, it suffers from one of the main limitations of the GP paradigm: while this EA performs really well in the generation of structures, adapting them both by crossover and mutation, the learning of the numeric values of the constants considered in the encoded structure – which are generated by the implementation program when the GP starts – can only be altered by mutation. This way, good trees solving the problem can be discarded by the selection procedure as the parameters involved in them are not well adjusted.

Hence, in the problem of extended Boolean query learning, the GP algorithm is able to find the positive, or negative, terms expressing the user's needs and to appropriately combine them by means of the logical operators AND and OR. However, it is very difficult for the algorithm to obtain the term weights, which constitutes a significant drawback due to their importance in the query.

Several solutions have been proposed for this GP problem. On the one hand, one can use a local search algorithm to learn the coefficients associated to each tree in the population [23]. On the other hand, the GA-P paradigm [21], an hybrid algorithm combining traditional GAs with the GP technique, can be considered to concurrently evolve the tree and the coefficients used in them, both of them encoded in the individual being adapted. Thus, each population member will involve both a value string and an expression. While the GP part of the GA-P evolves the expressions, the GA part concurrently evolves the coefficients used in them.

Most of the GA-P elements are the same as in either of the traditional genetic techniques. The GA-P and GP make selection and child generation similarly, except that the GA-P structure requires separate crossover and mutation operators for the expression and coefficient string components. Mutation and crossover rates for the coefficient string (using traditional GA methods) are independent from the rates for the expression part (using standard GP methods).

Taking the previous aspect into account, in [8, 9] we introduced a new IQBE technique for learning extended Boolean queries based on the GA-P technique. The different components of this algorithm are reviewed as follows.

Coding Scheme

When considering a GA-P to learn fuzzy queries, the expressional part (GP part) encodes the query composition – terms and logical operators – and the coefficient string (GA part) represents the term weights, as shown in Figure 2. In our case, a real coding scheme is considered for the GA part.

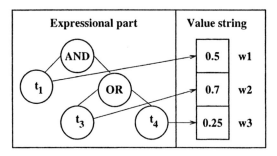

Fig. 2. GA-P individual representing the fuzzy query $0.5\ t_1\ AND\ (0.7\ t_3\ OR\ 0.25\ t_4)$

Selection Scheme

As in Kraft et al.'s algorithm, it is based on the classical generational scheme, together with the elitist selection. The intermediate population is created from the current one by means of Tournament selection.

Genetic Operators

A real-coded crossover operator – the BLX-α [15] – is considered for the GA parts. This operator generates an offspring, $C = (c_1, \dots, c_n)$, from two parents, $X = (x_1, \dots, x_n)$ and $Y = (y_1, \dots, y_n)$, with c_i being a randomly (uniformly) chosen number from the interval $[min_i - I \cdot \alpha, max_i + I \cdot \alpha]$, where $max_i = max\{x_i, y_i\}$, $min_i = min\{x_i, y_i\}$, and with $I = max_i - min_i$ ($[min_i, max_i]$ is the interval where the $i - th$ gene is defined). In our case, $[min_i, max_i] = [0, 1]$ and the operator is always applied twice to obtain two offsprings.

On the other hand, *Michalewicz's non-uniform mutation operator*, introduced in the previous section, is considered to perform mutation in the GA part.

As regards the operators for the GP part, the usual GP crossover described in the previous section is used, while the two first GP mutation operators (a) and b)) considered by the Kraft et al.'s algorithm are employed with probability 0.5 each.

Generation of the Initial Population and Fitness Function

Both have the same definition as those in Kraft et al.'s proposal, introduced in the previous section.

4 A Pareto-based Multiobjective GA-P Algorithm to Derive Extended Boolean Queries

The Pareto-based multiobjective EA considered to be incorporated to the GA-P algorithm introduced in the previous section has been Fonseca and Fleming's MOGA [17], one of the classical, first generation multiobjective EAs. Up to our knowledge, the only previous contributions incorporating Pareto-based multiobjective techniques into a GP-based algorithm are those by Rodríguez-Vazquez et al. [27] and by Cordón et al. [10].

The selection scheme of MOGA is based on dividing the population in several ranked blocks and assigning a higher probability of selection to the blocks with a lower rank, taking into account that individuals in the same block will be equally preferable and thus will receive the same selection probability. The rank of an individual in the population (and consequently of his belonging block) will depend on the number of individuals dominating it.

Therefore, the selection scheme of our multiobjective GA-P (MOGA-P) algorithm involves the following four steps:

1. Each individual is assigned a rank equal to the number of individuals dominating it plus one (chromosomes encoding non-dominated solutions receive rank 1).
2. The population is increasingly sorted according to that rank.
3. Each individual is assigned a fitness value which depends on its ranking in the population. In this contribution, we consider the following assignment: $f(C_i) = \frac{1}{rank(C_i)}$.
4. The fitness assignment of each block (group of individuals with the same rank, i.e., which are non dominated among them) is averaged among them, so that all of them finally receive the same fitness value.

Once the final fitness values have been computed, a usual selection mechanism is applied. In this contribution we consider the tournament selection introduced in Section 3.1 with an appropriate choice of the tournament size t to induce diversity.

A said in Section 2.4, it is well known that the MOGA selection scheme can cause a large selection pressure that might produce premature convergence. Fonseca and Fleming considered this issue and suggested to use a niching method to appropriately distribute the population in the Pareto [17].

This way, in this paper we apply niching in the objective space, in order to allow the algorithm to obtain a well-distributed set of queries with a different tradeoff between precision and recall, i.e., our initial aim. To do so, we make use of the usual Euclidean metric in order to measure the closeness between two different queries on the objective space.

Once a valid metric has been selected, it is easy to apply sharing by using the classical Goldberg and Richardson's sharing function [18]:

$$F(C_i) = \frac{f(C_i)}{\sum_{j=1}^{M} Sh(d(C_i, C_j))} \quad ; \quad Sh(d) = \begin{cases} 1 - (\frac{d}{\sigma_{share}})^\gamma, & \text{if } d < \sigma_{share} \\ 0, & \text{otherwise} \end{cases}$$

with σ_{share} being the niche radius.

5 Experiments Developed and Analysis of Results

This section is devoted to test the performance of the MOGA-P IQBE algorithm for FIRSs introduced in this paper. To do so, both the Kraft et al.'s algorithm and our MOGA-P proposal have been run to generate extended Boolean queries for the well known *Cranfield* collection, composed of 1400 documents about Aeronautics.

These 1400 documents have been automatically indexed by first extracting the non-stop words, and then applying a stemming algorithm, thus obtaining a total number of 3857 different indexing terms, and then using the normalized IDF scheme (see Section 2.2) to generate the term weights in the document representations.

Among the 225 queries associated to the Cranfield collection, we have selected those presenting 20 or more relevant documents in order to have enough chances to show the performance advantage of one algorithm over the other. The resulting seven queries and the number of relevant documents associated to them are showed in Table 1.

Table 1. Cranfield queries with 20 or more relevant documents

#query	#relevant documents
1	29
2	25
23	33
73	21
157	40
220	20
225	25

As said, apart from our MOGA-P, Kraft et al.'s IQBE process has been run on the seven selected queries for comparison purposes. In order to make this comparison fair, both algorithms have been provided with the same parameter values (see Table 2) and have been run a different number of times with different initializations till the same fixed number of fitness function evaluations have been performed.

As seen, the expressional part has been limited to 20 nodes in every case and populations are composed of 1600 individuals (the high value for this parameter is because it is well known that GP requires large population sizes to achieve good performance). For the sake of simplicity, only experiments

Table 2. Common parameter values considered

Parameter	Decision
Fitness function	F_2
Population size	1600
Number of evaluations	100000
Tournament size	16
Kraft et al.'s Crossover and Mutation probability	0.8, 0.2
Kraft et al.'s α, β weighting coefficientes in F_2	(0.8,1.2), (1,1), (1.2,0.8)
MOGA-P GA and GP Crossover probability	0.8, 0.8
MOGA-P GA and GP Mutation probability	0.2, 0.2
Expression part limited to	20 nodes
Retrieval threshold σ	0.1

not considering the use of the NOT operator are reported (as done in [24]). MOGA-P has been run ten different times for each query in order to check the robustness of the algorithm. The sharing function parameter γ takes value 2 and the niche radius σ_{share} has been experimentally set to 0.1.

On the other hand, Kraft et al.'s technique has been run considering three different values for the parameters α and β weighting, respectively, the precision and recall measures in the F_2 fitness function, in order to check the performance of the single-objective algorithm when being guided to different zones of the Pareto front. Three different runs have been done for each combination of values, thus making a total of nine runs for each query. All the runs have been performed in a 350 MHz Pentium II computer with 64 MB of RAM[6].

The results obtained by Kraft et al.'s algorithm are showed in Tables 3 and 4 respectively, with the average results being showed on the former table and the best ones on the latter. In the first table, $\#q$ stands for the corresponding query number, (α, β) for the values associated to the fitness function weigthing parameters, Sz for the average of the generated queries size and σ_{Sz} for its standard deviation, and P and R for the average of the precision and recall values (respectively, σ_P and σ_R for their standard deviations). The columns of the other table stand for the same items as well as Run for the number of the run where the reported result was derived, $\#rt$ for the number of documents retrieved by the query, and $\#rr$ for the number of relevant documents retrieved.

Tables 5 and 6 show several statistics corresponding to our multiobjective proposal. The former table collects several data about the composition of the ten Pareto sets generated for each query, always showing the averaged value and its standard deviation. From left to right, the columns contain the number of non-dominated solutions obtained ($\#p$), the number of different objective vectors (i.e., precision-recall pairs) existing among them ($\#dp$), and the values

---

[6] Kraft et al.'s algorithm spends more or less 13 minutes whilst MOGA-P approximately takes 15 minutes.

Table 3. Average results obtained by the single-objective Kraft et al.'s IQBE algorithm

#q	(α, β)	Sz	σ_{Sz}	P	σ_P	R	σ_R
1	1.2,0.8	19	0.0	1.0	0.0	0.3103	0.0912
	1.0,1.0	19	0.0	1.0	0.0	0.2873	0.0796
	0.8,1.2	15	2.0	0.0207	0.0	1.0	0.0
2	1.2,0.8	19	0.0	1.0	0.0	0.3866	0.0230
	1.0,1.0	19	0.0	1.0	0.0	0.3333	0.0461
	0.8,1.2	18.33	1.1547	0.01785	0.0	1.0	0.0
23	1.2,0.8	19	0.0	1.0	0.0	0.3232	0.1224
	1.0,1.0	19	0.0	1.0	0.0	0.2121	0.0909
	0.8,1.2	15	4.0	0.0235	0.0	1.0	0.0
73	1.2,0.8	19	0.0	1.0	0.0	0.5079	0.0727
	1.0,1.0	19	0.0	1.0	0.0	0.5714	0.0824
	0.8,1.2	18.33	1.1547	0.015	0.0	1.0	0.0
157	1.2,0.8	19	0.0	1.0	0.0	0.2583	0.0144
	1.0,1.0	19	0.0	1.0	0.0	0.175	0.05
	0.8,1.2	16.33	2.3094	0.0285	0.0	1.0	0.0
220	1.2,0.8	19	0.0	1.0	0.0	0.5166	0.0763
	1.0,1.0	19	0.0	1.0	0.0	0.5	0.05
	0.8,1.2	18.33	1.1547	0.0446	0.0525	1.0	0
225	1.2,0.8	19	0.0	1.0	0.0	0.44	0.04
	1.0,1.0	19	0.0	1.0	0.0	0.4266	0.0923
	0.8,1.2	16.33	3.0550	0.0178	0.0	1.0	0.0

of two of the usual multiobjective EA metrics $\mathcal{M}_2^*$ and $\mathcal{M}_3^*$ [36][7], all of them followed by their respective standard deviation values.

As regards the later metrics, $\mathcal{M}_2^* \in [0, \#p]$ measures the distribution of the objective vectors of the $\#p$ non-dominated solutions in the derived Pareto set Y' (i.e., the diversity of the solutions found) by means of the following expression:

$$\mathcal{M}_2^*(Y') = \frac{1}{|Y'-1|} \sum_{p' \in Y'} |\{q' \in Y'; \|p'-q'\|^* > \sigma^*\}|$$

with $\sigma^* > 0$ being a neighborhood parameter, and $\|\cdot\|$ being a distance metric. In this contribution, σ^* is set to σ_{share}, the niche radius considered, and $\|\cdot\|$ is the Euclidean distance. Of course, the higher the value of the measure, the better the distribution of the solutions within the Pareto front in the objective space.

On the other hand, $\mathcal{M}_3^*$ estimates the range to which the Pareto front spreads out in the objective values as follows:

---

[7] We should note that a third metric is proposed in that paper, $\mathcal{M}_1^*$, that can not be used in this contribution as it needs from the real Pareto front in order to be computed, which is not known in this case.

Table 4. Best results obtained by the single-objective Kraft et al.'s IQBE algorithm

#q	(α, β)	Run	Sz	P	R	#rr/#rt
	(1.2,0.8)	2	19	1.0	0.4137	12/12
1	(1.0,1.0)	3	19	1.0	0.3793	11/11
	(0.8,1.2)	1,2,3	13	0.0207	1.0	29/1400
	(1.2,0.8)	1,2	19	1.0	0.4	10/10
2	(1.0,1.0)	1,3	19	1.0	0.36	9/9
	(0.8,1.2)	1,2,3	17	0.0178	1.0	25/1400
	(1.2,0.8)	1	19	1.0	0.4545	15/15
23	(1.0,1.0)	3	19	1.0	0.3030	10/10
	(0.8,1.2)	1,2,3	11	0.0235	1.0	33/1400
	(1.2,0.8)	3	19	1.0	0.5714	12/12
73	(1.0,1.0)	1,2	19	1.0	0.6190	13/13
	(0.8,1.2)	1,2,3	17	0.015	1.0	21/1400
	(1.2,0.8)	1	19	1.0	0.275	11/11
157	(1.0,1.0)	2	19	1.0	0.225	9/9
	(0.8,1.2)	1,2,3	15	0.0285	1.0	40/1400
	(1.2,0.8)	1	19	1.0	0.6	12/12
220	(1.0,1.0)	2	19	1.0	0.55	11/11
	(0.8,1.2)	3	19	0.1052	1.0	20/190
	(1.2,0.8)	1	19	1.0	0.48	12/12
225	(1.0,1.0)	1,2	19	1.0	0.48	12/12
	(0.8,1.2)	1,2,3	13	0.0178	1.0	25/1400

$$\mathcal{M}_3^*(Y') = \sqrt{\sum_{i=1}^{n} \max\{\|p' - q'\|^*; p', q' \in Y'\}}$$

Since our problem is composed of just two objectives, it is equal to the distance among the objective vectors of the two outer solutions (hence, the maximum possible value is $\sqrt{2} = 1.4142$). Again, the higher the value, the larger the extent of the Pareto being covered.

Besides, two queries are selected from each Pareto set, the ones with maximum precision and maximum recall, respectively, and their averaged results are collected in Table 6.

In view of these results, the performance of our proposal is very significant. On the one hand, it overcomes the single-objective Kraft et al.'s algorithm in all cases but one (the best precision value obtained for query 23) as the results obtained by the latter when considering typical values for the weighted combination are dominated by the solutions in the Pareto front of the former, both in precision and recall. It seems that the diversity induced by the Pareto-based selection and the niching scheme make MOGA-P converge to better space zones. Notice the bad results obtained by the single-objective Kraft et al.'s process when giving more importance to the recall objective $((\alpha, \beta) = (0.8, 1.2)$ combination), as in every case but one (query 220), a query retrieving

Table 5. Statistics of the Pareto sets obtained by the proposed MOGA-P IQBE algorithm

#q	#p	$\sigma_{\#p}$	#dp	$\sigma_{\#dp}$	M_2^*	$\sigma_{M_2^*}$	M_3^*	$\sigma_{M_3^*}$
1	350.2	27.155	15.5	1.125	136.125	10.178	1.035	0.022
2	333.2	20.039	11.2	0.881	129.594	7.913	0.964	0.026
23	406.4	20.794	20.6	1.151	149.522	4.956	1.067	0.026
73	277.0	17.474	8.5	0.652	104.339	8.843	0.901	0.025
157	421.1	19.433	22.0	1.086	161.545	5.145	1.137	0.018
220	269.7	15.190	7.1	0.263	103.616	6.930	0.729	0.026
225	312.8	18.280	12.8	0.903	123.589	6.887	1.030	0.013

Table 6. Extreme solutions in the Pareto sets obtained by the proposed MOGA-P IQBE algorithm

#q	Best Precision						Best Recall					
	Sz	σ_{Sz}	P	σ_P	R	σ_R	Sz	σ_{Sz}	P	σ_P	R	σ_R
1	19.0	0.0	1.0	0.0	0.452	0.017	18.8	0.190	0.123	0.018	1.0	0.0
2	19.0	0.0	1.0	0.0	0.464	0.014	18.2	0.580	0.200	0.025	1.0	0.0
23	19.0	0.0	1.0	0.0	0.415	0.017	17.0	0.800	0.110	0.028	1.0	0.0
73	18.8	0.190	1.0	0.0	0.676	0.018	18.8	0.190	0.161	0.023	1.0	0.0
157	18.6	0.379	1.0	0.0	0.333	0.022	18.0	0.949	0.083	0.022	1.0	0.0
220	19.0	0.0	1.0	0.0	0.680	0.016	19.0	0.0	0.346	0.025	1.0	0.0
225	19.0	0.0	1.0	0.0	0.484	0.012	19.0	0.0	0.109	0.010	1.0	0.0

the whole documentary base is obtained, thus showing the bad convergence of the algorithm when considering these weighting factor values.

On the other hand, the main aim of this paper has been clearly fulfilled since the Pareto fronts obtained are very well distributed, as demonstrated by the high number of solutions included in them and the high values in the M_2 and M_3^* metrics. Maybe the only problem found is that the number of solutions presenting different precision-recall values (different objective value arrays) can be a little bit low with respect to the large number of solutions in the Pareto set. We think that this can be solved by considering a second generation Pareto-based approach, making use of a elitist population of non-dominated solutions.

As an example, Figures 3 and 4 show the Pareto fronts obtained for queries 1 and 157, representing the precision values in the X axis and the recall ones on the Y axis. As done in [36], the Pareto sets obtained in the ten runs performed for each query were put together, and the dominated solutions where removed from the unified set before plotting the curves.

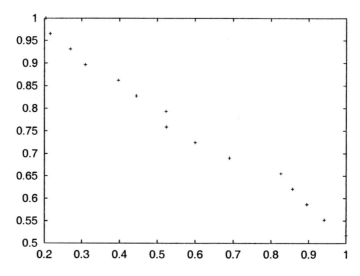

Fig. 3. Pareto front obtained for query 1

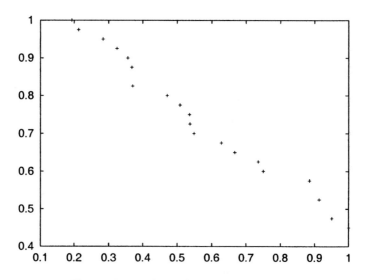

Fig. 4. Pareto front obtained for query 157

6 Concluding Remarks

The automatic derivation of extended Boolean queries has been considered by
incorporating the MOGA Pareto-based multiobjective evolutionary approach
to an existing GA-P-based IQBE proposal. The proposed approach has performed appropriately in seven queries of the well known Cranfield collection

in terms of absolute retrieval performance and of the quality of the obtained Pareto sets.

In our opinion, many different future works arise from this preliminary study. On the one hand, as it has been mentioned before, more advanced Pareto-based multiobjective EA schemes, such as those second generation elitist ones considering an auxiliary population to better cover the Pareto front (see in Section 2.4), can be incorporated to the basic GA-P algorithm in order to improve the performance of the multiobjective EA proposed. On the other hand, preference information of the user on the kind of queries to be derived can be included in the Pareto-based selection scheme in the form of a goal vector whose values are adapted during the evolutionary process [17]. Moreover, a training-test validation procedure can be considered to test the real-world applicability of the proposed IQBE algorithm. Finally, and more generically, Pareto-based multiobjective evolutionary optimization can be applied either to the automatic derivation of queries for other kinds of IR models or to other IR problems being solved by EAs [7], thus benefiting from the potential of these techniques in the problem solving.

References

1. Bäck T (1996) Evolutionary algorithms in theory and practice. Oxford University Press.
2. Baeza-Yates R, Ribeiro-Neto, B (1999) Modern information retrieval. Addison-Wesley.
3. Bordogna G, Carrara P, Pasi G (1995) Fuzzy approaches to extend Boolean information retrieval. In: Bosc P, Kacprzyk J (eds) Fuzziness in database management systems. Physica-Verlag, pp. 231–274.
4. Chankong V, Haimes Y Y (1983) Multiobjective decision making theory and methodology. North-Holland.
5. Chen H, Shankarananrayanan G, She L, Iyer A (1998) Journal of the American Society for Information Science 49(8):693–705.
6. Coello C A, Van Veldhuizen D A, Lamant G B (2002) Evolutionary algorithms for solving multi-objective problems. Kluwer Academic Publishers.
7. Cordón O, Moya F, Zarco C (April, 1999) A brief study on the application of genetic algorithms to information retrieval (in spanish). In: Proc. Fourth International Society for Knowledge Organization (ISKO) Conference (EOCON-SID'99), Granada, Spain, pp. 179–186.
8. Cordón O, Moya F, Zarco C (September, 1999) Learning queries for a fuzzy information retrieval system by means of GA-P techniques. In: Proc. EUSFLAT-ESTYLF Joint Conference, Palma de Mallorca, Spain, pp. 335–338.
9. Cordón O, Moya F, Zarco C (2000) Mathware & Soft Computing 7(2-3):309–322.
10. Cordón O, Moya F, Zarco C (2002) Soft Computing 6(5):308-319.
11. Cordón O, Herrera-Viedma E, Luque M (September, 2002) Evolutionary learning of Boolean queries by multiobjective genetic programming. In: Proc. Seventh Parallel Problem Solving from Nature (PPSN-VII) International Conference, Granada, Spain, LNCS 2439. Springer, pp. 710-719.

12. Cross V (1994) Journal of Intelligent Information Systems 3:29–56.
13. Deb K, Goldberg D E (1989) An investigation of niche and species formation in genetic function optimization. In: Proc. Third International Conference on Genetic Algorithms (ICGA'89), Hillsdale, USA, pp. 42-50.
14. Deb K (2001) Multi-objective optimization using evolutionary algorithms. Wiley.
15. Eshelman L J, Schaffer J D (1993) Real-coded genetic algorithms and interval-schemata. In: Whitley L D (ed) Foundations of Genetic Algorithms 2, Morgan Kaufmann, pp. 187–202.
16. Fogel D B (1991) System identification trough simulated evolution. A machine learning approach. Ginn Press, USA.
17. Fonseca C M, Fleming P J (July, 1993) Genetic algorithms for multiobjective optimization: Formulation, discussion and generalization. In: Proc. Fifth International Conference on Genetic Algorithms (ICGA'93), San Mateo, CA, pp. 416–423.
18. Goldberg D E, Richardson J (1987) Genetic algorithms with sharing for multimodal function optimization. In: Proc. Second International Conference on Genetic Algorithms (ICGA'87), Hillsdale, USA, pp. 41–49.
19. Gordon M, Pathak P (1999) Information Processing and Management 35(2):141–180.
20. Herrera-Viedma E (2001) Journal of the American Society for Information Science 52(6):460–475.
21. Howard L, D'Angelo D (1995) IEEE Expert: 11–15.
22. Ide E (1971) New experiments in relevance feedback. In: Salton G. (ed) The SMART Retrieval System. Prentice Hall, pp. 337–354.
23. Koza J (1992) Genetic programming. On the programming of computers by means of natural selection. The MIT Press.
24. Kraft D H, Petry F E, Buckles B P, Sadasivan T (1997) Genetic algorithms for query optimization in information retrieval: relevance feedback. In: Sanchez E, Shibata T, Zadeh L A (eds) Genetic algorithms and fuzzy logic systems. World Scientific, pp. 155–173.
25. Michalewicz Z (1996) Genetic algorithms + data structures = evolution programs. Springer-Verlag.
26. Mitchel T M (1997) Machine learning. McGraw-Hill.
27. Rodríguez-Vazquez K, Fonseca C M, Fleming P J (July, 1997) Multiobjective genetic programming: A nonlinear system identification application. In: Late Breaking Papers at the Genetic Programming 1997 Conference, Stanford, USA, pp. 207–212.
28. Salton G, McGill M J (1989) Introduction to modern information retrieval. McGraw-Hill.
29. Sanchez E (1989) Information Systems 14(6):455–464.
30. Sánchez L, Couso I, Corrales J A (2001) Information Sciences 136(1-4):175-191.
31. Schaffer J D (1985) Multiple objective optimization with vector evaluated genetic algorithms. In: Genetic algorithms and their applications. Proc. of the First International Conference on Genetic Algorithms, pp. 93-100.
32. Schwefel H-P (1995) Evolution and optimum seeking. Sixth-Generation Computer Technology Series. John Wiley and Sons.
33. Smith M P, Smith M (1997) Journal of Information Science 23(6):423–431.
34. van Rijsbergen C J (1979) Information retrieval (2nd edition). Butterworth.

35. Zadeh L A (1965) Information and Control 8:338–353.
36. Zitzler E, Deb K, Thiele L (2000) Evolutionary Computation 8(2):173–195.

An Approximate Querying Environment for XML Data

Ernesto Damiani[1], Nico Lavarini[2], Barbara Oliboni[2], and Letizia Tanca[2]

[1] Università di Milano, Dipartimento di Tecnologie dell'Informazione
Via Bramante 65, 26013 Crema, Italy
`edamiani@crema.unimi.it`
[2] Politecnico di Milano, Dipartimento di Elettronica e Informazione
Via Ponzio 1, 20100 Milano, Italy
`oliboni,tanca@elet.polimi.it`

Summary. Often, XML information from heterogeneous sources carries the same semantics but turns out to be structured in different ways. In this Chapter we extend our previous work about *blind* pattern-based XML querying by presenting an approximate querying technique capable to locate and extract information dealing flexibly with differences in structure and tag vocabulary.

Our approach relies on representing XML documents as graphs, whose edges are weighted at different levels of granularity. We describe a content-insensitive *automatic* weighting technique taking into account various characteristics of each edge, generating a separate weight according to each characteristic, and then aggregating these values in a single arc-weight. Then, a threshold-based pruning of unimportant edges is performed, allowing us to retain only the most useful information, improving search performance. Edge weighting enables us to tackle the problem of estimating similarity between XML nodes by means of *type-aware* functions. In particular, we deal with simple types such as URIs and dates, where the usual string-based comparison hardly captures the similarity between values.

Finally, weighting allows us to introduce fuzzy operators inside the query itself, using *linguistic variables* and *fuzzy quantification* operators. Linguistic variables are used to extend patterns using fuzzy substitutes expressed in natural language (e.g. "*many*", "*tall*", etc), rather than numerical values. Fuzzy quantification is used for requesting an approximate *number of matchings* for a part of a pattern. In other words, the user can require the result to contain a (fuzzy) number of subtrees of a particular kind, rooted at a particular node.

1 Introduction

The widespread use of the *eXtensible Mark-up Language* (XML) as a general format for representing, exchanging and publishing information on the Web has raised the need of new techniques for XML information processing.

Namely, novel XML query and search techniques are needed to be able to locate, extract and organize information from heterogeneous XML documents, dealing flexibly with differences in structure and tag vocabulary.

Our research is aimed at a flexible search and processing technique, capable to extract relevant information from a (possibly huge) set of heterogeneous XML documents or data flows. We intend to deal with XML data coming from a number of different sources, each employing its own mark-up; this corresponds to a high degree of variability about the documents' structure and tag vocabulary. Such a scenario is particularly common when a user search involves XML data sources situated across organizational boundaries. Indeed, when multiple data sources are managed by different organizations, the same kind of information is usually structured in different ways; moreover, it is normal for the user not to know all data variations in advance.

In our approach, the user simply provides a *XML pattern*, i.e. a partially specified XML (sub)tree. Then, the XML documents of the target dataset are scanned, their graphs[3] are searched and the XML fragments matching the pattern are located and sorted according to their similarity to the pattern. Of course, the notion of subgraph matching needs to be carefully tuned in order to maximize recall while avoiding a massive decrease in precision. As a first step, in [12] we described a fuzzy technique for extending the structure of target document graphs and weighting their arcs (obtaining a *weighted XML graph*), in order to increase pattern recall. The problem of matching the pattern to the extended XML graph structure was first tackled in [12] by means of a *fuzzy sub-graph matching* technique, allowing for a user-controlled degree of approximate matching between the pattern and the candidate fragments of the target document. Subsequently, several other approaches to approximate XML querying and searching were proposed, many of them providing some sort of result ranking [19, 21, 24].

In order to develop our research on this topic, in this Chapter we provide a short summary of our previous results in Section 2. Then, we deal with other fundamental aspects of the approximate querying problem, taking into account the role of XML content and attribute nodes as well as structure. Namely, in Section 3, we introduce a novel *automatic* edge-weighting technique aimed at obtaining a format for *weighted* XML graphs that takes into account content, while ensuring compatibility with the XML standard. Then, in Section 4 we deal with the problem of estimating type-based similarity between nodes and propose some type-aware functions to compute the similarity degree. In Section 5 we show how search patterns can be enriched by using fuzzy operators based on *linguistic variables* and *fuzzy quantification*,

---

[3] While XML documents' structure is indeed a tree w.r.t. element containment, it becomes a graph if explicit and/or implicit links are taken into account. In the remainder of the Chapter we shall use the term *tree* only when explicitly referring to element containment.

72

the latter allowing the user to introduce *fuzzy counting* in a query. At the end Section 6 draws the conclusions and Appendix 7 shows a worked-out example.

2 The framework

In this Section we recall our fuzzy technique for extending the structure of XML document trees and weighting their arcs. This technique was first proposed in [12] in order to increase pattern recall of XML searches, while preserving some notion of the target documents' hierarchical structure. A software environment based on this approach was presented in [14].

Our approach relies on modeling XML documents as labeled graphs and computing estimates of the *importance* of the information they provide in each node. Note that, though the use of attributes in XML documents should be confined to express data properties, XML documents found in practice very often carry data both in attributes and in PCDATA content.

To our purposes, these two kinds of data should be treated accordingly; thus, in our framework, all the attribute values of a node are considered as subelements of the given element. This helps searching, since many XML document writers use attributes as standard content and our recall is not limited by the designer's project decisions. The result of such extension is a *fuzzy labeled graph*[4] [5]. Then, we proceed to selectively extend the fuzzy graph's structure by computing its fuzzy closure; since each closure arc corresponds to a path in the original graph, it gets a weight computed by aggregating the weights of the arcs and nodes of the path by means of a context-dependent *t-norm*. In this framework, a query can be readily represented as a *graph pattern*; in order to compute the "degree of similarity" between a pattern and a document we use a subgraph matching procedure and present a *ranked list* containing the query result ordered by degree of similarity w.r.t. the user query.

The approach of [12] relies upon three basic steps:

1. **Assignment of weights** to the target document content (nodes and arcs) on the basis of the document's topological structure, and tag repertoire. This step is carried out manually at document design time and highlights information considered important by the document designer, at the granularity of XML tags and attributes. We distinguish between *structure-related* and *tag-related* weighting techniques: the former attaches importance to the containment relationships that specify the spatial structure of the documents (not unlike weighting in image databases [15]), while the latter expresses the importance of XML elements and attributes content *per se*. In [12] edge-weighting was done manually, based on nodes depth. In this Chapter we shall discuss an *automatic* weighting technique (Section 3). Figure 1 shows a weighted XML document.

---

[4] i.e. a graph labeled on nodes (by XML element names), and weighted on arcs.

2. **Extension of the fuzzy labeled graph**. In this step, which can be automatically carried out at any phase of the document's life cycle, the fuzzy transitive closure of the weighted document graph is computed. At query execution time, the result is then tailored by performing a thresholding α-cut operation on the basis of a threshold parameter provided by the user, or computed by the system on the basis of the user's profile. Edges of the closure graph whose weight is below the α threshold are deleted. The output of this step is a new, tailored target graph (See Figure 2).

3. **Computation of a similarity matching** between the subgraphs of the tailored document and the query pattern. The user can choose among a suite of similarity matchings, going from strict (isomorphism-based) to loose (bisimulation-based). The choice of matching conveys the intended query semantics [20, 10]. While subgraph matching tolerates variations in XML documents' structure, it does not deal with XML content. Figure 3 shows a pattern which does not match the graph of Figure 2, since the requested date is incomplete. As we shall see, this situation can be dealt with by using approximate *type-driven* matching techniques.

3 Automatic weighting

Document weighting is fundamental in our framework, as it allows graph edges (and nodes) to carry further information beyond their content. This auxiliary information provides a rating of the node importance that is exploited by our threshold-based cutting of "unimportant" edges. Importance estimates are useful only inasmuch they are reliable, and the topic of automatically weighting all the arcs of a document graph is quite complex. Of course, we are fully aware that a "perfect" weighting technique should be fully content-cognizant, and that "correct" importance attribution can be made only by hand. However, experience shows that even manual weighting has a non-negligible error rate, since the weighting task becomes harder as the depth and branching of the target XML tree increases. Since documents' number and size usually make manual weighting unpractical, an automatic topology-related weighting procedure should be attempted anyway. For instance, a method based on a (normalized, inverse-of) distance from the root can be conceived, based upon the hypothesis (borrowed from object-oriented design) that generality (and thus, probably, importance) grows when getting closer to the root. However, we must also take into account the fact that in XML documents most content is stored inside terminal elements, possibly at the maximum depth. This means that a reliable weighting technique should not allow for easily threshold-cutting edges leading to content-filled elements. As a result, the following factors have been chosen as the basis of our automatic arc weighting method:

Depth: The "closeness" to the root, that shows the generality of the concepts associated to elements connected to the arc.

74

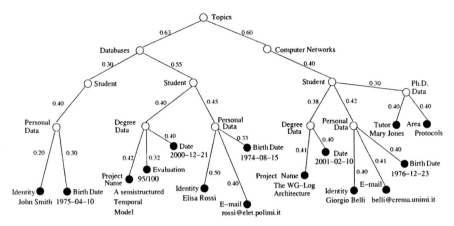

Fig. 1. A weighted XML document

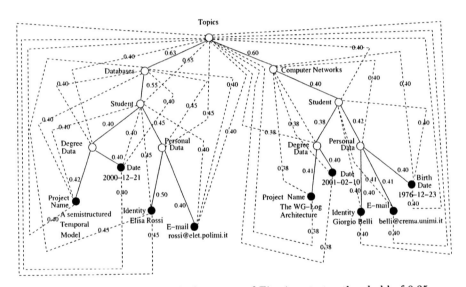

Fig. 2. Closure of the XML document of Fig. 1, cut at a threshold of 0.35.

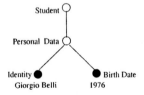

Fig. 3. A pattern

Content: The "amount" of PCDATA content, as an estimate of the fact that the arc leads to *real* data.

Fan-out: The number of nodes directly reachable from the destination node of the arc, to indicate the arc leads to a wide branching of concepts.

Tag name: Though content-insensitive, our technique could take into account the presence of particular tag names (possibly by using domain specific Thesauri), and increase weights of arcs incoming to nodes whose names are considered as "content bearers". This factor heavily depends on the reliability of the underlying dictionary; therefore we do not take it into account in this analysis.

Our automatic weighting technique takes into account separately the factors listed above, generates a value for each of them, and then aggregates these values within a single arc-weight.

Depth In order to mirror the fact that the containment relationship between XML elements causes generality to decrease as depth grows, we define a standard decreasing hyperbolic function (similar to the one we will show in Fig. 5(a)) that gives the lowest weights to the deepest nodes.

If $a = (n_1, n_2)$ is an arc then its weight according to depth is

$$w_d(a) = \frac{\alpha}{\alpha + \mathrm{depth}(n_1)}$$

where α can be easily tuned. Let us suppose, as an example, to have a tree with maximum depth 10. It is easy to see that, with $\alpha = 1$ the weights go from 1 to $1/11$, and with $\alpha = 10$ the weights go from 1 to $1/2$. The choice of α can also depend on the value of the maximum depth D. It is easy to show that, in this case, if $\alpha = D/k$ then the minimum weight is $1/(k+1)$.

Fan-out The importance of an arc is also given by the number of nodes it leads to. In fact, the fan-out of an arc (n_1, n_2) is defined as the number of elements contained in n_2. To express a value related to fan-out we use again a simple function.

If $F(a)$ is the fan-out of arc a, we can define

$$w_f(a) = \frac{F(a)}{F(a) + \beta}$$

with $\beta > 0$ to be tuned as needed. In this case, the function obtained is similar to that in Fig. 4(a). If we assume $\beta = 1$ we have that, if the fan-out of an arc is k, then its weight will be $k/(k+1)$, asymptotic to 1 as k grows.

Content The importance of leaf nodes tends to be underestimated by the weights just described, because leaf nodes are often deep inside the document, and have no children. This may sound counter-intuitive because leaf nodes are the main information bearers, and a threshold-cutting on them would eliminate potentially useful content. For this reason, we also weight nodes *based on the amount of content*. Since our weighting technique is content-insensitive, the only way we have to quantify the amount of information in

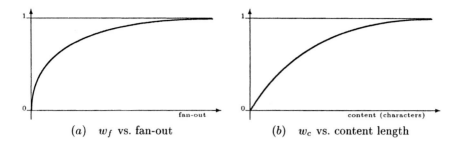

(a) w_f vs. fan-out   (b) w_c vs. content length

Fig. 4. Hyperbolic functions

a text node is to calculate the length of the corresponding PCDATA string. This means that, given an arc $a = (n_1, n_2)$ its weight (w.r.t. content) should be proportional to the length of the text contained in n_2.

If $C(a)$ is the text content of the destination node of a, then

$$w_c(a) = \frac{|C(a)|}{|C(a)| + \gamma}$$

where $|C(a)|$ is the length of $C(a)$, that can be expressed either in tokens (words) or in bytes (characters). As usual, a parameter γ can be used to tune the slope. Actually γ represents the length of the content for which $w_c = 0.5$.

Aggregation Up to this point we obtained a set of three parameters derived respectively from depth (w_d), fan-out (w_f) and content (w_c) that need to be combined in order to obtain the final weight. Often there is a trade-off between w_d and w_c, and we have to consider an aggregation which takes into account all the three elements. To this end, we employ a *ordered weighting operator* [22]: the weight of an edge a is computed as

$$w(a) = \alpha_1 w_d(a) + \alpha_2 w_c(a) + \alpha_3 w_f(a)$$

with $\alpha_1 + \alpha_2 + \alpha_3 = 1$.

4 Type-driven Matching

We are now ready to describe a method to compute type-based similarity between a content node of the pattern (n^p) and one of the document(n^d), taking into account their estimated type T. In particular, we will refer to URIs and dates, where string comparison hardly captures the real similarity between values.

4.1 URI Type

A node whose type has been estimated as URI can be matched with more precision than a simple string. Each URI can be roughly split into three distinct parts: a *host name*, a *path* and a *file name*[5].

`www.informatik.uni-freiburg.de`	`/~may/Mondial/`	`mondial-3.0.xml`
host name	*path*	*file name*

A partial match in each of these parts can be significant: suppose a URI u_1 in a query is matched against an element with value u_2 in a document, such that u_2 is found to be a URI (here we do not consider to deal with a document base located in a single website, but with a general document set, spread all over the Web).

- A *host name* match means that the two URIs share the common base origin, and thus a positive match can be returned.
  Note that in some cases a partial match can be found, e.g. between `www.microsoft.com` and `research.microsoft.com`[6].
- A *path* match might be useful when dealing with *mirrors*. Two URIs might thus indicate identical documents (given they share the same name) located on different mirrors of the same document base. In general, the longer the path (i.e. the deeper the document in the file system tree), the more it is likely to find a correct match; for example, if we find two `index.xml` documents, both with path `/docs/` we should return a very small match value.
- A *file name* match might be useful only if involved names are "uncommon" (think, for example, to `RussianRecipes2001.xml`), or when the match is combined with a host name and/or path match. A perfect match on a dummy "`index.html`" name should be considered meaningless.

As an example, suppose the contents of n^p and n^d to be respectively
`www.informatik.uni-freiburg.de/~may/Mondial/mondial-3.0.html` and
`www.uni-freiburg.de/research/may/Mondial/mondial-3.0.html`
The different parts are dealt with in the following way:

Host name A function f_T^h can be easily defined. We consider the host name as a list of dot-separated words, where the first element is the rightmost one. We also call $l_{\max}$ the length of the longer of the two URIs, and $l_{\min}$ the length of the shorter one. The similarity f_T^h between the two URIs can

---

[5] Actually, the analysis can be further refined by considering that a complete URI may contain other elements: the protocol specification, the TCP port number, some parameters or a destination anchor name, etc. (like, for instance, `http://www.google.com:80/search?q=xml&btnG=Search&hl=en`). We do not deal with all of them for the sake of conciseness.

[6] The "www" field could be considered non significant and thus excluded from the similarity computation (see also footnote 5).

78

be evaluated by aggregating the similarities between the single words, in the following way:

$$f_T^h(h_1, h_2) = \frac{1}{l_{\max}} \sum_{k=1}^{l_{\min}} \frac{1}{k} C(h_1(k), h_2(k))$$

where

$$C(a, b) = \begin{cases} 1 & \text{if } a = b \\ 0 & \text{if } a \neq b \end{cases}$$

Thus with `www.informatik.uni-freiburg.de` and `www.uni-freiburg.de` we have the following computation (with n_h we denote the substring of the URI n representing the host name, and analogously for the path n_p and the file name n_n):

$$f_T^h(n_h^p, n_h^d) = \frac{1}{4} \sum_{k=1}^{3} \frac{1}{k} \times C(n_h^p(k), n_h^d(k))$$
$$= \frac{1}{4} \times [(1 \times 1) + (\frac{1}{2} \times 1) + (\frac{1}{3} \times 0) + (\frac{1}{4} \times 0)] = 0.375$$

Actually, the URIs have the same part in position $k = 1$ and $k = 2$.

As other examples, with `www.google.com` and `www.polimi.it` we have $l_{\max} = 3$, $l_{\min} = 3$ and the computation is:

$$f_T^h(n_h^p, n_h^d) = \frac{1}{3} \sum_{k=1}^{3} \frac{1}{k} \times C(n_h^p(k), n_h^d(k))$$
$$= \frac{1}{3} \times [(1 \times 0) + (\frac{1}{2} \times 0) + (\frac{1}{3} \times 1)] = 0.111$$

and with `citeseer.nj.nec.com` and `www.elet.polimi.it` we have $l_{\max} = 4$, $l_{\min} = 4$ and the computation is:

$$f_T^h(n_h^p, n_h^d) = \frac{1}{4} \sum_{k=1}^{4} \frac{1}{k} \times C(n_h^p(k), n_h^d(k))$$
$$= \frac{1}{4} \times [(1 \times 0) + (\frac{1}{2} \times 0) + (\frac{1}{3} \times 0) + (\frac{1}{4} \times 0)] = 0$$

Actually `www.elet.polimi.it` and `citeseer.nj.nec.com` are different in all parts.

Path In order to define a suitable function f_T^p in this case, we simply observe that two paths can be considered as matching if they share a common significant (w.r.t. length) suffix. Given two paths p_1 and p_2, let us define p_s as the longest string which is a suffix both of p_1 and p_2. Thus, indicating with $|s|$ the length of the string s, we can define

$$f_T^p(p_1, p_2) = \begin{cases} \frac{|p_s|}{\max(|p_1|, |p_2|)} & \text{if } |p_s| \neq 0 \\ 0 & \text{otherwise} \end{cases}$$

As far as our example is concerned, with $n_p^p = $ ~`may/Mondial` and $n_p^d = $ `research/may/Mondial`, we have $p_s = $ "may/Mondial", and $f_T^p(n_p^p, n_p^d) = 11/20 = 0.55$. Note that we stripped the leading and trailing slashes, to avoid a certain though meaningless match on the final character.

File name We consider a match on the file name significant if (i) the match is perfect; (ii) the name is not an element of a tiny "dictionary of common web-document names" (CND), which contains, for instance, `index.*`, `home.*`, `default.*`, etc. Furthermore, the match is the more significant the longer the filename is. Thus, we can define our function as follows:

$$f_T^n(n_1, n_2) = \begin{cases} \frac{|n_1|}{|n_1|+1} & \text{if } n_1 = n_2 \text{ and } n_1 \notin \text{CND} \\ 0 & \text{otherwise} \end{cases}$$

Recalling our example, with the common file name `mondial-3.0.html`, and considering it an uncommon name, we have $f_T^n(n_n^p, n_n^d) = 16/17 = 0.941$.

Now, a simple aggregation operator suffices to compute the match value between two URIs. Namely, we use again an *ordered weighting operator* [22]:

$$f_T(n^p, n^d) = \alpha_1 f_T^h(n_h^p, n_h^d) + \alpha_2 f_T^p(n_p^p, n_p^d) + \alpha_3 f_T^n(n_n^p, n_n^d)$$

with $\alpha_1 + \alpha_2 + \alpha_3 = 1$. Practical experience led us to set $\alpha_1 = 0.6$, $\alpha_2 = 0.3$ and $\alpha_3 = 0.1$. In the case of our example, the resulting match value is:

$$f_T(n_{p_i}, n_{d_j}) = 0.6 \times 0.375 + 0.3 \times 0.55 + 0.1 \times 0.941 = 0.484$$

Note that the use of a standard metrics to compare URIs, e.g. the Edit Distance [3], would not provide good results in our setting. Such metrics is a general-purpose string comparison, which returns a distance between strings, without taking into account any kind of semantics of URI strings; furthermore, it behaves incorrectly (to our purposes) in the case of simple string variants, like for example anagrams.

4.2 "Date" Type

Suppose now that the two nodes n_{p_i} and n_{d_j} are both dates. Advantages in using a specific `Date` type include the possibility of quantifying the match in a subtler way than when considering simple strings.

Our method for matching dates considers the match value as inversely proportional to the distance between the dates, measured in days. This difference can be automatically calculated by simple library functions, included in most high-level programming languages.

Our technique for match value computation exploits a hyperbolic, decreasing function of the distance k between the dates (See Fig. 5(a)), computing the value as $v_d = \frac{\alpha}{\alpha+k}$. The parameter α can be tuned to get a customized function slope, in order to obtain a faster or slower convergence to zero w.r.t. the distance.

Note that all values of α clearly return 1 for two identical dates, and that the function value decreases as the distance grows. For instance, a $\alpha = 1/2$ returns $v_d = 1/3$ for a 1-day distance, $v_d = 1/15$ for a 1-week distance, and $v_d = 1/61$ for a 1-month distance. The result value could be amended in

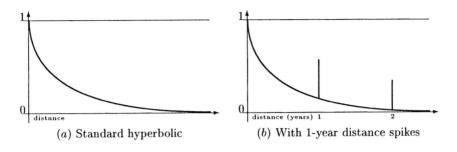

(a) Standard hyperbolic   (b) With 1-year distance spikes

Fig. 5. Function *Date*

particular cases: suppose for instance that we find exactly a 1-year difference between two dates. This value could be considered as more significant than others, for it could represent the date of a periodic event (See Fig. 5(b)). In this case we use a simple technique. Given two dates, and their match value v_d obtained as explained above, if their distance is found to be exactly y years (i.e. day and month are the same), the correct match value v_f is calculated as

$$v_f = v_d + \frac{1 - v_d}{1 + y} \quad = \frac{1 + yv_d}{1 + y}$$

i.e. adding to the estimated value an amount that is inversely proportional to the distance in years. The resulting function shows, in correspondence of exactly an integer number of years y, some spikes, whose height is inversely proportional to y.

As an example, suppose a quite liberal $\alpha = 10$. In this case, the simple match value v_d between 1998-05-25[7] and 1999-05-25 would be very small (0.027), but the final value is

$$v_f = 0.027 + \frac{0.973}{2} = 0.5135$$

The match value v_d between 1998-05-25 and 2000-05-25 would instead be 0.0135, and the final value is thus

$$v_f = 0.0135 + \frac{0.9865}{3} = 0.3423$$

Incomplete dates Sometimes XML documents contain incomplete dates. This means that, either because the writer did not know the precise date, or because some part of it is intended to be understood, the day or the year are not included. We assume that the month is almost never skipped if not together with the day. Thus, we often find documents with strings like

---

[7] To avoid mismatches, we use a ISO 8601-compliant date notation, as proposed in note [25]. See also http://www.cl.cam.ac.uk/~mgk25/iso-time.html.

"December 2000", or "June, 12". In this case the correct matching procedure should assume that missing data (after a suitable format conversion) have to be considered as correct as possible, e.g. a comparison between 12-06 and 2000-12-06 should return a difference of 0 days. Trouble comes when the two dates are, for instance, 1998-12-28 and 01-03. If we consider the missing year as 1998 we obtain a difference of 359 days, but the two dates could be much closer. A correct evaluation should consider this, computing the distance between two dates (of which at least one is year-lacking) as $v_d = \min[k, 365 - k]$ where k is the distance in days, calculated from the first date to the second, without taking into account the year. For the example above, the result should be $v_d = \min(359, 6) = 6$, which actually is the distance between 12-28 and 01-03 (of the following year). If the day is lacking in one date, we consider the day of the month nearest to the other date. This means that in a match between 2000-07 and 2000-08-03 the first date is considered to be 2000-07-31, but if the second date were 2000-05-03 the first would be considered to be 2000-07-01. For instance, if we need the distance between 2000-08-03 and 2000-07, the result will be

$$v_d = \min\left(d(2000\text{-}07\text{-}01, 2000\text{-}08\text{-}03), d(2000\text{-}07\text{-}31, 2000\text{-}08\text{-}03)\right)$$

which returns the value $v_d = 3$. By means of this technique the pattern of Fig. 3 can be successfully matched against the graph of Fig. 2, since the incomplete date 1976 matches with 1976-12-23. In some cases, dates contain also the time of the day. Our technique deals efficiently with this feature; we calculate the difference in days as a floating point value, which contains also the possible time, and thus can evaluate the function in a precise point.

5 Linguistic variables and fuzzy quantification

So far we have dealt with the problem of XML search by matching crisp patterns against fuzzy graphs, discussing enhancements based on type-aware similarity. Now, we are ready to introduce some fuzzy elements in our query patterns as well. Namely, we shall deal with

Linguistic variables which allow for requesting values the query result must contain using, rather than numerical values, fuzzy elements, expressed in natural language (e.g. "*many*", "*tall*", etc.);

Fuzzy quantification which requests an *approximate cardinality* of the query result.

Both the use of linguistic variables defined on attributes' domains and fuzzy quantification are well-known techniques and have been used since long to deal with uncertainty in fuzzy relational databases [16].

In order to understand the need for their extension to the case of XML data, consider the document in Fig. 6, representing a pretty uncommon individual. This document suggests that it should be possible to enrich patterns both by linguistic variables based upon the value of an element and by fuzzy quantification on the number of elements (or subelements, at arbitrary depth) contained in a specified element. Namely, we can:

1. apply linguistic variables to the tag `<age>`, once recognized it has a numerical value, and thus search for old people (in which `<age>` has a high value).
2. use fuzzy quantification to count the number of tags `<wife>` of each `<person>`, extracting people who have a *"high"* number of wives.

```
<person id="123">
        <name>Jack</name>
        <age>98</age>
        <wife id="1">Anne</wife>
        <wife id="2">Bess</wife>
        <wife id="3">Clara</wife>
        <wife id="4">Dana</wife>
        <wife id="5">Emma</wife>
</person>
```

Fig. 6. Document fragment

To further clarify this distinction, consider Fig. 7. Query (a) extracts only graphs in which the element F's value is HIGH. Query (b), instead, should match graphs in which element E contains MANY subtrees matching with T_2.

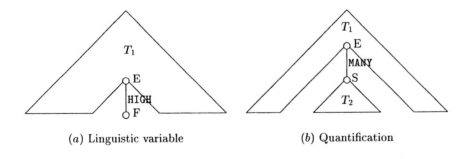

(a) Linguistic variable   (b) Quantification

Fig. 7. Linguistic variables and fuzzy quantification in XML querying

In both cases, some XML-specific problems arise. In the former case, we need to recognize that the element value is numeric, which may lead to mistakes when no schema information is available. In the latter case, the problem is even more difficult: we have to verify that the nodes matching S (with their subtrees) should be considered together, i.e. belong of the same type.

In general this is not a simple task, especially when dealing with complex subtrees.

Recalling the example in Fig. 6, the problem amounts to finding, among all the elements contained in <person>, which ones should be included in the count. However, dealing with this problem is worthwile, as fuzzy quantification turns out to be particularly relevant to XML querying. When querying structured data, it is always possible to explicitly count the number of values by using of aggregations like SQL's COUNT. Thus, there appears to be little need to query whether they are MANY, when we can know their exact number. On the other hand, in semistructured data languages do not usually provide counting primitives, and thus a fuzzy way to group values is indeed very useful. As we shall see, our approach provides full support for fuzzy counting, though in some cases the choice of a suitable aggregation parameter will be left to the user.

5.1 Linguistic variables

In our approach, queries of this kind are modeled by allowing the use of keywords such as HIGH, MEDIUM and LOW as labels for the pattern arcs (See Fig. 8(b)).

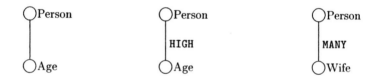

(a) Simple query   (b) With a linguistic variable   (c) With a fuzzy quantifier

Fig. 8. Sample queries

Informally, the semantics of this query is as follows; the graph pattern has the usual meaning (in the Figure, to extract Person elements which own an Age subelement), but it is required (for HIGH) to return those document fragments for which the destination node of the arc has a *high* value (analogously for MEDIUM and LOW). In other words, the destination node of the arc has to be an element containing text, possibly including subelements. Its value has to belong to a *numeric* type. If these conditions are satisfied, we can decide whether a value is high or not through the use of *fuzzy membership functions* [22] (See Fig. 9).

These functions are the fuzzy correspondent of set characteristic functions, and are defined by two parameters: μ is the average value of the dataset, possibly calculated as a sample mean; ξ is a gauge for the "fuzziness" of the function[8]. Low values of ξ lead to crisp characteristic functions, while high ones tend to blur values, showing little difference for a value variation.

Fuzzy membership functions are weakly monotonic, i.e. $\mathcal{F}_h(x)$ must be increasing, $\mathcal{F}_l(x)$ must be decreasing, and $\mathcal{F}_m(x)$ must be increasing for $x < \mu$, and decreasing for $x > \mu$. Moreover, both $\mathcal{F}_l(\mu) \neq 0$ and $\mathcal{F}_h(\mu) \neq 0$, to avoid empty results if $\sigma = 0$, i.e. the value is constant throughout the document base.

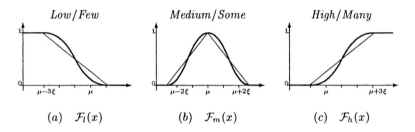

(a) $\mathcal{F}_l(x)$ (b) $\mathcal{F}_m(x)$ (c) $\mathcal{F}_h(x)$

Fig. 9. Fuzzy membership functions

The use of these functions is indeed simple: when, searching for HIGH values of an element, we find a value of x, we compute $k = \mathcal{F}_h(x)$, expressing *how much x is* HIGH. Since these functions can be well approximated by polygonals (as shown by thin lines in Fig. 9), their computation is usually fast.

Once matching subgraphs have been identified they must be sorted taking into account the fact that unlike the case of crisp patterns, both the membership value of the element v and the match value m (i.e. the similarity estimation) must now be taken into account. Fuzzy text retrieval systems often use product-based aggregation for match values ($r = mv$) [22].

In our case, however, the two values carry very different semantics and we may need to tune the aggregation operator. To this purpose, we can easily apply a *dampening* exponent $k \in [0, 1]$ to the value whose effect we want to reduce. If we apply dampening to m, we get a *value-oriented aggregation*, in which we consider graph structures as correctly matching; on the other hand, applying dampening to v reduces the incidence of the value on the result, obtaining thus a *structure-oriented aggregation* (See Table 1).

In each case, dampening becomes stronger as k decreases; actually, if $k = 1$ we again have the standard product, and if $k = 0$ the dampening effect is maximal, and the result takes into no account the value of the dampened parameter.

In the following, we shall use the standard product-based aggregation.

---

[8] ξ can possibly depend on the variance, e.g. $\xi = \sigma$. In this case, the shape of the function depends on the value distribution of the dataset.

Politics	Aggregation	
Standard aggregation	$r = mv$	
Value-oriented	$r = m^k v$	$(k \in [0,1])$
Structure-oriented	$r = mv^k$	$(k \in [0,1])$

Table 1. Value aggregation

Ranking	1	2	3	4	...
Result	Person 1 Age 50	Person 0.9 Age 40	Person 0.7 Age 60	Person 0.5 Age 65	...
Match value	1	0.9	0.7	0.5	...

Fig. 10. Sample result of Query in Fig. 8(a)

As an example, if we look for old people (**persons** with a **HIGH age**) and find a man of 60 (suppose $\mathcal{F}_h(60) = 1$) with match value 0.8, and a woman of 50 (suppose $\mathcal{F}_h(50) = 0.9$) with value 1, we consider the latter as the more significant, because their aggregate values are $1 \times 0.8 = 0.8$ for the former and $0.9 \times 1 = 0.9$ for the latter. What we do is thus multiplying the function *value* found by its *likelihood*.

As another simple example, consider the small query of Fig. 8(a). It requires to find **persons** who own a **age** subelement. Suppose now the answer to this query be the list of fragments of Fig. 10. In this case the results are ranked according to their match value, which is possibly less than 1, since edges weights can vary being obtained by computing the fuzzy graph closure.

Suppose now to modify the query to that of Fig. 8(b), which just requires to find old people, and to submit it to the same document base. The result set we obtain is a subset of the previous one, whose elements belong to the set of **high** ones with positive value. Anyway, this result set is now sorted in a different way (See Fig. 11).

In this case, for each result fragment f_i (whose element value is v_i), with its graph match value m_i, we computed its relative match value as $r_i = m_i \mathcal{F}_h(v_i)$, considered as the gauge of the importance. Thus, results are ordered according to the decreasing value of r_i.

Note that the ordering of these results is quite different from the one we would have obtained by considering only the value of the **age** subelement. For example, the element with the largest age 65 is at the third place, because of its low reliability.

Ranking	1	2	3	4	...
Result (f_i)	◯Person 1 ◯Age 50	◯Person 0.7 ◯Age 60	◯Person 0.5 ◯Age 65	◯Person 0.9 ◯Age 40	...
m_i	1	0.7	0.5	0.9	...
v_i	50	60	65	40	...
$\mathcal{F}_h(v_i)$	0.7	0.9	1	0.3	...
r_i	**0.7**	**0.63**	**0.5**	**0.27**	...

Fig. 11. Sample result of Query in Fig. 8(b)

5.2 Fuzzy quantification

The second possibility for the use of fuzzy operators in the queries (as shown in Fig. 6) is in "fuzzy counting" the number of subtrees of a given node, which can be FEW, SOME or MANY.

The user specifies this quantification by enriching the search pattern with the keywords MANY, SOME and FEW, as arc labels.

Recalling Fig. 6, a query made by two nodes person and wife, connected by an arc labeled MANY would require to extract all person nodes with many wife children elements (Fig. 8(c)).

To further clarify the difference between fuzzy quantification and the use of linguistic variables, recall Fig. 7 where (a) and (b) are two generic queries whose syntax is similar but semantics is completely different.

Let us first analyze query (a). To compute its result, we find all the matches for the query graph T_1, including nodes E and F. At this point, the ordering of the result presentation is modified, taking into account the value of $\mathcal{F}_h(F.text)$ of each found fragment, and multiplying it by the match value of all the graph.

As far as query (b) is concerned, instead, the query resolution procedure is different. We need to find the graphs similar to T_1 which have (contained in element E) *many* subtrees similar to T_2. As an example, the document graph fragment in Fig. 12 shows a generic tree with such characteristics. In this case the tree T_1' matches with pattern T_1 of the query (without taking into account the children of E), and the node E' owns many (namely 3) children nodes which are roots of subtrees T_2', T_2'', T_2''' matching with T_2.

First, suppose the match between T_1 and T_1' produces a value v_1.

At this point, we have to count the number of matches for the subtree T_2: in this case the three subtrees T_2', T_2'' and T_2''', resp. match T_2 with values v_2', v_2'' and v_2'''. The fuzzy number of subtrees is then computed as

$$v_2 = v_2' + v_2'' + v_2''' \leq 3$$

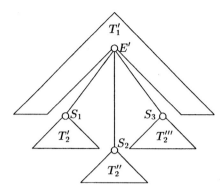

Fig. 12. Sample document fragment

Note that this "fuzzy adding" works properly only if the edge-aggregation operator T is chosen for the fuzzy graph closure such that $\forall i \in 1 \ldots k$ we have that $T(w_1, \ldots, w_k) \leq w_i$ [13]. These v_2 subtrees are thus MANY with value $v_h = \mathcal{F}_h(v_2)$.

At this point the match value of the complete query can be obtained by aggregating the match value found for T'_1 and the *high* value for the "number of T_2s" (See Table 1). Using standard product-based aggregation, we obtain a final match value $v = v_1 v_h$ between the pattern in Fig. 7(*b*) and the fragment in Fig. 12.

In general, a matching containing a quantification of this type will require several match operations. Consider Fig. 7(*b*). First, we have to perform a simple fuzzy match using the graph without the subtree rooted at S. At this point we must take into account only those results for which there exists a positive match for node E, and call each of these matching nodes in a result E^i. For each result, we have now to match the quantified part. For each E^i we search subtrees T_2^i, rooted at E^i and matching T_2. Only in this way it is possible to ensure a correct fuzzy quantification of elements. Actually, this kind of matching amounts to applying a crisp match to the document graph; the difference is that, in the crisp case, a document like the one in Fig. 12 would produce exactly three matches: for the subtrees $(T'_1 \cup T'_2)$, $(T'_1 \cup T''_2)$ and $(T'_1 \cup T'''_2)$.

6 Conclusions

In this Chapter, we presented an approach to approximate search in heterogeneous XML data sources, relying on representing XML documents as graphs whose edges are weighted at different levels of granularity. An *automatic* weighting technique was described, allowing for threshold-based pruning of unimportant edges. Then, we showed how the computation of similarity between XML nodes can be enhanced by means of simple *type-aware* functions.

As an example, we dealt with elementary types such as URIs and dates, where the usual string-based comparison hardly captures the similarity between values. Finally, we discussed how our query patterns can be enriched with fuzzy operators, using *linguistic variables* and *fuzzy quantification*. The latter technique turns out to be particularly relevant to semi-structured information querying. Our approach has been tested using a sample XML database with encouraging results. We have already implemented a fully-functional prototype of a structure-based XML search engine, and we are currently investigating the design of such an engine providing the scalability required by modern Web-based applications, as well as adding some of the new features described in this Chapter to our prototype.

References

1. Bordogna G, Lucarella D, Pasi G (1994) A Fuzzy Object Oriented Data Model. IEEE International Conference on Fuzzy Systems, vol. 1, pp. 313–317.
2. Bouchon-Meunier B, Rifqi M, Bothorel S (1996) Towards General Measures of Comparison of Objects. Fuzzy Sets and Systems, vol. 84.
3. Burchard P (2000) Cut-Paste Metrics: A More Natural Approach to Approximate String Matching. http://www.pobox.com/~burchard/bio/fuzzymatch/.
4. Ceri S, Bonifati A (2000) Comparison of XML Query Languages. SIGMOD Record 29(1).
5. Chan KP, Cheung YS (1992) Fuzzy Attribute Graph with Applications to Character Recognition. IEEE Trans. on Systems, Man and Cybernetics, 22(1).
6. Cohen R, Di Battista G, Kanevsky A, Tamassia R (1993) Reinventing the Wheel: An Optimal Data Structure for Connectivity Queries. Proc. of ACM-TOC Symp. on the Theory of Computing. S.Diego CA, (USA).
7. Cohen S, Kogan Y, Nutt W, Sagiv Y, Serebrenik A (2000) EquiX: Easy Querying in XML Databases. WebDB (Informal Proceedings).
8. Comai S, Damiani E, Posenato R, Tanca L (1998) A Schema-Based Approach to Modeling and Querying WWW Data. In Cristiansen H, ed., Proceedings of Flexible Query Answering Systems (FQAS '98), Roskilde (Denmark), Lecture Notes in Artificial Intelligence 1495, Springer.
9. Ceri S, Comai S, Damiani E, Fraternali P, Paraboschi S, Tanca L (1999) XML-GL: a Graphical Language for Querying and Restructuring XML Documents. Computer Networks, Vol. 31, pp. 1171–1187.
10. Cortesi A, Dovier A, Quintarelli E, Tanca L (1998) Operational and Abstract Semantics of a Query Language for Semi-Structured Information. To appear in Theoretical Computer Science (preliminary version in Proc. of DDLP'98, pp. 127–139. GMD Report 22).
11. Damiani E, Fugini MG, Bellettini C (1999) A Hierarchy Aware Approach to Faceted Classification of Object-Oriented Components. ACM Transactions on Software Engineering Methodologies, vol. 8, n. 3 pp. 215–262.
12. Damiani E, Tanca L (2000) Blind Queries to XML Data. Proceedings of DEXA 2000, London, UK, September 4–8. Lecture Notes in Computer Science, Vol. 1873, pp. 345–356, Springer.

13. Damiani E, Tanca L, Arcelli Fontana F (2000) Fuzzy XML Queries via Context-based Choice of Aggregations. Kybernetika n. 16 vol. 4.
14. Damiani E, Lavarini N, Marrara S, Oliboni B, Pasini D, Tanca L, Viviani G (2002) The APPROXML Tool Demonstration. VIII Conference on Extending Database Technology (EDBT 2002). LNCS 2287, pp. 753–755.
15. Del Bimbo A, Vicario E (1998) Using Weighted Spatial Relatioship in Retrieval By Visual Content. Proc. IEEE Workshop on Content Based Access of Images, Santa Barbara CA (USA).
16. Dubois D, Prade H, Testemale C (1984) Generalizing Database Relational Algebra for the Treatment of Uncertain Information and Vague Queries. Information Sciences Vol. 34 n. 2, pp. 115–143.
17. Dubois D, Esteva F, Garcia P, Godo L, Lopez de Mantaras R, Prade H (1998) Fuzzy Set Modelling in Case-Based Reasoning. Int. Jour. of Intelligent Systems 13(1).
18. Dubois D, Prade H, Sedes F (1999) Fuzzy Logic Techniques in Multimedia Database Querying: A Preliminary Investigations of the Potentials. In Meersman R, Tari Z and Stevens S, eds., Database Semantics: Semantic Issues in Multimedia Systems, Kluwer Academic Publisher.
19. Fuhr N, Grojohann K (2001) XIRQL: A Query Language for Information Retrieval in XML Documents. ACM SIGIR Conference on Research and Development in Information Retrieval.
20. Gold S, Rangarajan A (1996) A Graduated Assignment Algorithm for Graph Matching. IEEE Trans. on Pattern Analysis and Machine Intelligence, 18(2).
21. Hayshi Y, Tomita J, Kikui G (2000) Searching Text-rich XML Documents with Relevance Ranking. ACM SIGIR 2000 Workshop on XML and Information Retrieval.
22. Klir GJ, Folger TA (1988) Fuzzy sets, uncertainty, and information. Prentice Hall, Englewood Cliffs, USA.
23. May W (1999) Information extraction and integration with FLORID: The MONDIAL case study. Technical Report 131, Universität Freiburg, Institut für Informatik. http://www.informatik.uni-freiburg.de/~may/Mondial/.
24. Schlieder T, Meuss H (2000) Result Ranking for Structured Queries against XML Documents. DELOS Workshop: Information Seeking, Searching and Querying in Digital Libraries.
25. W3C. Date and Time Formats — Note. Acknowledged submission to W3C from Reuters Limited. http://www.w3.org/TR/NOTE-datetime/.
26. W3C. Extensible Stylesheet Language (XSL) Version 1.0. October 2000. http://www.w3C.org/TR/xsl/.
27. W3C. Extensible Markup Language (XML) 1.0. Feb. 1998. http://www.w3C.org/TR/REC-xml/.
28. W3C. Resource Description Framework (RDF): Model and Syntax Specification. W3C Recommendation 22 February 1999. http://www.w3.org/TR/REC-rdf-syntax/.
29. W3C. XSL Transformations (XSLT) Version 1.0. W3C Recommendation 16 November 1999. http://www.w3.org/TR/xslt/.
30. W3C. Namespaces in XML. W3C Recommendation 14 January 1999. http://www.w3.org/TR/1999/REC-xml-names-19990114/.

7 Appendix: A Worked-out Example

We have tested our environment [14] on several document bases. For the sake of conciseness, here we will show the results obtained by querying a XML World Geographical Atlas, known as MONDIAL [23]. The MONDIAL database has been compiled from geographical Web data sources, like the CIA World Factbook (`http://www.cia.gov/cia/publications/factbook/`) and many others. It is widely acknowledged that its complexity and size make MONDIAL a good benchmark for XML processing[9]. In the remainder of the Section, we shall discuss a part of the result set of some fuzzy queries (See Fig. 13).

As explained in Section 2, we deal with attributes like subelements. Note that, in this case (See Fig. 14 for a fragment of the MONDIAL database), attributes are indeed used as "content bearers" thus, if we searched only among elements, we would lose much of the database content (e.g., query Q_4 would return no result).

Many of these queries, like for example Q_1, have been submitted to test the *likelihood* of the answers, which are well-known[10].

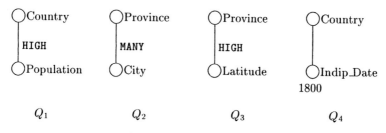

Fig. 13. Example queries

7.1 Querying results

Query Q_1 requires to extract countries with a `HIGH` value of population. A structure match is found for every country in the database; the mean value and standard deviation of population are thus calculated, i.e. $\mu = 24,953,358$ and $\sigma = 105,038,765$. Result includes the most populated countries.

`--- Result ranking: 1 - match value 1.0 ---`

```
<country>
    <name> China </name>
    <population> 1216310400 </population>
    <total_area> 9596960 </total_area>
    <government> Communist state </government>
</country>
-------------------------------------------------
--- Result ranking: 2 - match value 0.999 ---
<country>
    <name> India </name>
    <population> 952107712 </population>
```

---

[9] The MONDIAL database is available in several versions. The one we used is at `http://www.informatik.uni-freiburg.de/~may/Mondial/mondial-3.0.xml`.

[10] Only top-ranked results for the queries of Fig. 13 are shown. XML resulting fragments are heavily *pruned* of unimportant subelements.

```
<?xml version="1.0" encoding="UTF-8"?>
<mondial id="result">
    <continent id="f0_114" name="Europe"/>
    <continent id="f0_118" name="Asia"/>
    ...
    <country id="f0_392" name="Sweden" capital="f0_1617"
     population="8900954" datacode="SW"
     total_area="449964" population_growth="0.56"
     infant_mortality="4.5" gdp_agri="2"
     gdp_total="177300" inflation="2.6"
     government="constitutional monarchy" gdp_ind="27"
     gdp_serv="71" car_code="S">
        <province id="f0_18122" name="ivsborg"
         country="f0_392" capital="f0_5537"
         population="444259" area="11395">
            <city id="f0_5537" country="f0_392"
             province="f0_18122">
                <name id="f0_5538">
                    Vanersborg
                </name>
                <population id="f0_52012" year="87">
                    35804
                </population>
                <longitude id="f0_51243"/>
                <latitude id="f0_51244"/>
            </city>
        </province>
        ...
    </country>
    ...
    <organization id="f0_29347" name="Nordic Council"
     abbrev="NC" established="1952-03-16">
        <members id="f0_47021" country="f0_188"/>
        <members id="f0_47022" country="f0_203"/>
        ...
    </organization>
    ...
    <island id="f0_37741" name="Mallorca" area="3618">
        <located id="f0_52553"/>
    </island>
    ...
    <river id="f0_38177" length="3531" name="Volga">
        <located id="f0_52729"/>
        <located id="f0_52730"/>
        ...
    </river>
    ...
    <lake id="f0_39376" name="Barrage de Mbakaou">
        <located id="f0_53121"/>
    </lake>
</mondial>
```

Fig. 14. A fragment of MONDIAL database

```
    <total_area> 3287590 </total_area>
    <government> federal republic </government>
</country>
-----------------------------------------
--- Result ranking: 3  -  match value 0.909 ---
<country>
    <name> USA </name> <name> states </name>
    <name> United States </name>
    <population> 266476272 </population>
    <total_area> 9372610 </total_area>
    <government> federal republic </government>
</country>
-----------------------------------------
--- Result ranking: 4  -  match value 0.895 ---
```

```
<country>
    <name> Indonesia </name>
    <population> 206611600 </populatic
    <total_area> 1919440 </total_area>
    <government> republic </government
</country>
-----------------------------------------
--- Result ranking: 5  -  match value
<country>
    <name> Brazil </name>
    <population> 162661216 </populatic
    <total_area> 8511965 </total_area>
    <government> federal republic </gc
</country>
```

```
--- Result ranking: 6  -  match value 0.833 ---
<country>
     <name> Russia </name>
     <population> 148178480 </population>
     <total_area> 17075200 </total_area>
     <government> federation </government>
</country>
```

Query Q_2 requires all the provinces including MANY cities as subelements. Since in the database several city elements appear as a country subelements, instead of a province one, recall is not total. In spite of this, the result is correct. The mean value is very small, since most of the provinces have a unique city subelement (their capital).

```
-- Result: 1 (#city 53)  - match value 0.950 --
<province>
     <name> Sao Paulo </name>
     <city>
          <name> Sao Paulo </name>
          <population> 9811776 </population>
     </city>
     <city>
          <name> Guarulhos </name>
          <population> 972766 </population>
     </city>
     <city>
          <name> Campinas </name>
          <population> 907996 </population>
     </city>
     ...
     ...
     <city>
          <name> Botucatu </name>
          <population> 100826 </population>
     </city>
     <city>
          <name> Ribeirao Pires </name>
          <population> 100335 </population>
     </city>
</province>
--------------------------------------------
-- Result: 2 (#city 51)  - match value 0.944 --
<province>
     <name> California </name>
     <city>
          <name> Los Angeles </name>
          <population> 3553638 </population>
     </city>
     <city>
          <name> San Diego </name>
          <population> 1171121 </population>
     </city>
     <city>
          <name> San Jose </name>
          <population> 838744 </population>
     </city>
     <city>
          <name> San Francisco </name>
          <population> 735315 </population>
     </city>
     ...
     ...
     <city>
          <name> Corona </name>
          <population> 100208 </population>
     </city>
     <city>
          <name> Concord </name>
          <population> 114850 </population>
     </city>
</province>
--------------------------------------------
-- Result: 3 (#city 31)  - match value 0.889 --
<province>
     <name> Nordrhein Westfalen </name>
     <city>
          <name> Dusseldorf </name>
          <population> 572638 </population>
     </city>
     <city>
          <name> Koln </name>
```

```
     <name> Cologne </name>
     <population> 963817 </population>
</city>
<city>
     <name> Essen </name>
     <population> 617955 </population>
</city>
<city>
     <name> Dortmund </name>
     <population> 600918 </population>
</city>
...
...
<city>
     <name> Witten </name>
     <population> 105423 </population>
</city>
<city>
     <name> Bergisch Gladbach </name>
     <population> 105122 </population>
</city>
</province>
--------------------------------------------
-- Result: 4 (#city 27)  - match value 0.850 --
<province>
     <name> Shandong </name>
     <city>
          <name> Jinan </name>
          <population> 2320000 </population>
     </city>
     <city>
          <name> Zibo </name>
          <population> 2460000 </population>
     </city>
     <city>
          <name> Qingdao </name>
          <population> 2060000 </population>
     </city>
     <city>
          <name> Yantai </name>
          <population> 452127 </population>
     </city>
     ...
     ...
     <city>
          <name> Jiaonan </name>
          <population> 121397 </population>
     </city>
     <city>
          <name> Zhucheng </name>
          <population> 102134 </population>
     </city>
</province>
```

Query Q_3 asks for the provinces with a HIGH latitude. Note that matchings are found throughout the database only by means of closure arcs, since the elements province do not include any latitude subelement or attribute.

```
--- Result ranking: 1  -  match value 0.490 ---
<province>
     <name> Arkhangelskaya oblast </name>
     <city>
          <longitude> 40.5333 </longitude>
          <latitude> 64.55 </latitude>
          <name> Arkhangelsk </name>
          <population> 374000 </population>
     </city>
</province>
--------------------------------------------
--- Result ranking: 2  -  match value 0.475 ---
<province>
     <name> Northwest Territories </name>
     <city>
          <longitude> -114.45 </longitude>
          <latitude> 62.4667 </latitude>
          <name> Yellowknife </name>
          <population> 8256 </population>
     </city>
</province>
--------------------------------------------
--- Result ranking: 3  -  match value 0.455 ---
<province>
     <name> Alaska </name>
     <city>
          <longitude> -150.017 </longitude>
```

```
        <latitude> 61.1667 </latitude>
        <name> Anchorage </name>
        <population> 250505 </population>
    </city>
</province>
-----------------------------------------------
--- Result ranking: 4  -  match value 0.437 ---
<province>
    <name> Uusimaa </name>
    <city>
        <longitude> 24.95 </longitude>
        <latitude> 60.1667</latitude>
        <name> Helsinki </name>
        <population> 487428 </population>
    </city>
</province>
-----------------------------------------------
--- Result ranking: 5  -  match value 0.430 ---
<province>
    <name> Oslo </name>
    <city>
        <longitude> 10.7333 </longitude>
        <latitude> 59.9333 </latitude>
        <name> Oslo </name>
        <population> 449337 </population>
    </city>
</province>
```

Query Q_4 requires the country which have an independence date equal to 1800. As we showed in Sect. 4.2, a fuzzy match value can be calculated also for incomplete dates.

```
--- Result ranking: 1  -  match value 0.909
<country>
    <name> United Kingdom </name>
    <indep_date> 1801-01-01 </indep_date>
</country>
-----------------------------------------------
--- Result ranking: 2  -  match value 0.007
<country>
    <name> Haiti </name>
    <indep_date> 1804-01-01 </indep_date>
</country>
-----------------------------------------------
--- Result ranking: 3  -  match value 0.003
<country>
    <name> Colombia </name>
    <indep_date> 1810-07-20</indep_date>
</country>
-----------------------------------------------
--- Result ranking: 4  -  match value 0.003
<country>
    <name> Mexico </name>
    <indep_date> 1810-09-16 </indep_date>
</country>
-----------------------------------------------
--- Result ranking: 5  -  match value 0.003
<country>
    <name> Chile </name>
    <indep_date> 1810-09-18 </indep_date>
</country>
-----------------------------------------------
--- Result ranking: 6  -  match value 0.002
<country>
    <name> Paraguay </name>
    <indep_date> 1811-05-14 </indep_date>
</country>
```

Information Gathering on the Internet Using a Distributed Intelligent Agent Model with Multi-Granular Linguistic Information

F. Herrera[1], E. Herrera-Viedma[1], L. Martínez[2], and C. Porcel[1]

[1] Dept. of Computer Science and Artificial Intelligence.
University of Granada, 18071 - Granada, Spain
`herrera,viedma,@decsai.ugr.es,cporcel@invest.ugr.es`
[2] Dept. of Computer Science.
University of Jaén, 23071 - Jaén, Spain
`martin@ujaen.es`

Summary. Information gathering in Internet is a complex activity. Find the appropriate information, required for the users, on the World Wide Web is not a simple task. Then, Internet users need tools to assist them to obtain the information required. One possibility consists of using distributed intelligent agents in the information gathering process that help the users to cope with the mass of content available on the World Wide Web.

The communication between users and agents is very important to the information gathering process be successful. The great variety of representations of the information in Internet is the main obstacle to this communication. The use of the linguistic information provides a more flexibility in the communication among agents and between agents and users. In this paper, we propose a distributed intelligent model for gathering information on the Internet, where the agents and users may communicate among them using a multi-granular linguistic model. This model provides a greater flexibility and several advantages in the user-system interaction.

Keywords: Internet, information retrieval, intelligent agents, computing with words, linguistic modelling.

1 Introduction

Information gathering on Internet is a very important, widely studied and hotly debated topic. The exponential increase in Web sites and Web documents is contributing to that Internet users not being able to find the information they seek in a simple and timely manner. There are many publicly available search engines, but users are not necessarily satisfied with the different formats for inputting queries, speed of retrieval, presentation formats of the retrieval results, and quality of retrieved information. Therefore, users are in need of tools to help them cope with the mass of content available on the World Wide Web [17, 18].

A solution consists in to assist Internet users in information gathering processes by means of distributed intelligent agents in order to find the fittest information to their information needs [3, 9, 25, 28, 32]. Several proposals about intelligent software agents have been emerging in the recent last years to improve different tasks related to *networking* among them the Information Retrieval. But the lack of connection and communication among agents have lead to a decrease in the quality and suitability of the retrieved information besides the efficiency of the system in the recovering and filtering task. The great variety of representations and evaluations of the information in the Internet is the main obstacle to this communication, and the problem becomes more noticeable when the user takes part in the process. The complexity of all these processes reveals the need of more flexibility in the communication among agents and between agents and the user [32, 33]. For this purpose, several approaches related to mechanisms to introduce and handle flexible information through linguistic information have been proposed both at levels of agents and users [6, 7, 31]. In such approaches as the user queries as the relevance degrees of retrieved documents are assessed using the same linguistic labels with the same semantics. However, it is obvious that both concepts are different and have a different interpretation. Therefore, it seems reasonable and necessary to assess them with different linguistic label sets, i.e., by using multi-granular linguistic assessments [11].

In this paper, we present a distributed intelligent agent model for gathering information on the Internet where the communication among the agents of different levels and among the agents and users is carried out by using a multi-granular linguistic modelling. We assume that in the agent system the user queries, the satisfaction degrees of user queries, and the relevance degrees of retrieved documents are assessed using different linguistic domains or label sets with different granularity. To do so, we use hierarchical linguistic contexts [14] based on the linguistic 2-tuple computational model [12] as representation base of the multi-granular linguistic information. In such a way, we achieve the following advantages: i) the retrieval process is endowed with a higher flexibility, ii) the expressiveness of agent system in the system-user interaction is improved and iii) the processes of computing with words (CW) are made without loss of information and therefore, with more precision.

This paper is structured as follows. In Section 2 we present a short review of the fuzzy linguistic approach, of the 2-tuple fuzzy linguistic representation model and of the hierarchical linguistic contexts. Section 3 shows the structure of the distributed intelligent agent model which uses the multi-granular linguistic model for information gathering. Section 4 presents an example for illustrating the proposal. Finally some conclusions are pointed out.

2 Linguistic Modelling

In this section we present the tools that allow us to apply the linguistic modelling in the distributed intelligent agent model.

2.1 Fuzzy Linguistic Approach

Many aspects of day-to-day activities are evaluated by means of imprecise and fuzzy qualitative values. As was pointed out in this may be arise for different reasons. There are some situations in which information may be unquantifiable due to its nature, and thus, it can be stated only in linguistic terms (e.g., when evaluating the "comfort" or "design" of a car, terms like "good", "fair", "poor" can be used). In other cases, precise quantitative information cannot be stated because either it is unavailable or the cost for its computation is too high and an "approximate value" can be tolerated (e.g., when evaluating the speed of a car, linguistic terms like "fast", "very fast", "slow" can be used instead of numeric values) [11].

The use of Fuzzy Sets Theory has given very good results for modelling qualitative information [34]. It is a technique that handles fuzzines and represents qualitative aspects as linguistic labels by means of "linguistic variables", that is, variables whose values are not numbers but words or sentences in a natural or artificial language. The linguistic approach is used in different fields, such as for example, "information retrieval" [2, 15, 16], "clinical diagnosis" [5], "decision making" [10], etc.

In any linguistic approach, an important parameter to determinate is the "granularity of uncertainty", i.e., the cardinality of the linguistic term set used to express the information. According to the uncertainty degree that an expert qualifying a phenomenon has on it, the linguistic term set chosen to provide his knowledge will have more or less terms. When different experts have different uncertainty degrees on the phenomenon, then several linguistic term sets with a different granularity of uncertainty are necessary (i.e. multigranular linguistic information) [11]. Typical values of cardinality used in the linguistic models are odd ones, such as 7 or 9, where the mid term represents an assessment of "approximately 0.5", and with the rest of the terms being placed symmetrically around it [1].

One possibility of generating the linguistic term set consists of directly supplying the term set by considering all terms distributed on a scale on which a total order is defined. For example, a set of seven terms S, could be given as follows:

$$S = \{s_0 = N, s_1 = VL, s_2 = L, s_3 = M, s_4 = H, s_5 = VH, s_6 = P\}.$$

Usually, in these cases, it is required that in the linguistic term set there exist:

1. A negation operator: $\text{Neg}(s_i) = s_j$ such that $j = g$-i (g+1 is the cardinality).

2. An order: $s_i \leq s_j \iff i \leq j$. Therefore, there exist the *min* and *max* operators.

The semantics of the linguistic terms is given by fuzzy numbers defined in the [0,1] interval. A way to characterize a fuzzy number is to use a representation based on parameters of its membership function [1]. The linguistic assessments given by the users are just approximate ones, some authors consider that linear trapezoidal membership functions are good enough to capture the vagueness of such linguistic assessments. The parametric representation is achieved by the 4-tuple (a, b, d, c), where b and d indicate the interval in which the membership value is 1, with a and c indicating the left and right limits of the definition domain of the trapezoidal membership function [1]. A particular case of this type of representation are the linguistic assessments whose membership functions are triangular, i.e., $b = d$, then we represent this type of membership functions by a 3-tuple (a, b, c). An example may be the following (Figure 1) :

$$N = (0, 0, .17) \quad VL = (0, .17, .33) \ L = (.17, .33, .5)$$
$$M = (.33, .5, .67) \ H = (.5, .67, .83) \ VH = (.67, .83, 1) \ P = (.83, 1, 1).$$

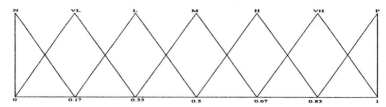

Fig. 1. A set of seven linguistic terms with its semantics

2.2 The 2-tuple Fuzzy Linguistic Representation Model Based on the Symbolic Translation

This model and its applications has been presented in [12, 13, 14], showing different advantages of this formalism for representing the linguistic information over classical models, such as:

1. The linguistic domain can be treated as continuous, while in the classical models it is treated as discrete.
2. The linguistic computational model based on linguistic 2-tuples carries out processes of computing with words easily and without loss of information.
3. The results of the processes of computing with words may be always expressed in the initial expression domain.
4. It is possible to aggregate multi-granular linguistic information easily.

Due to these advantages, we shall use this linguistic representation model to accomplish our objective: a higher flexibility, uniformity and precision in the retrieval process with multi-granular information.

The 2-tuple Fuzzy Linguistic Representation Model.

Let $S = \{s_0, ..., s_g\}$ be a linguistic term set, if a symbolic method aggregating linguistic information obtains a value $\beta \in [0, g]$, and $\beta \notin \{0, ..., g\}$ then an approximation function $(app_2(\cdot))$ is used to express the index of the result in S.

Definition 1. *Let β be the result of an aggregation of the indexes of a set of labels assessed in a linguistic term set S, i.e., the result of a symbolic aggregation operation. $\beta \in [0, g]$, being $g + 1$ the cardinality of S. Let $i = round(\beta)$ and $\alpha = \beta - i$ be two values, such that, $i \in [0, g]$ and $\alpha \in [-.5, .5)$ then α is called a Symbolic Translation.*

From this concept we shall develop a linguistic representation model which represents the linguistic information by means of 2-tuples (s_i, α_i), $s_i \in S$ and $\alpha_i \in [-.5, .5)$:

- s_i represents the linguistic label of the information, and
- α_i is a numerical value expressing the value of the translation from the original result β to the closest index label, i, in the linguistic term set (s_i), i.e., the Symbolic Translation.

This model defines a set of transformation functions between numeric values and 2-tuples.

Definition 2. *Let $S = \{s_0, ..., s_g\}$ be a linguistic term set and $\beta \in [0, g]$ a value representing the result of a symbolic aggregation operation, then the 2-tuple that expresses the equivalent information to β is obtained with the following function:*

$$\Delta : [0, g] \longrightarrow S \times [-0.5, 0.5)$$

$$\Delta(\beta) = (s_i, \alpha), \ with \ \begin{cases} s_i & i = round(\beta) \\ \alpha = \beta - i & \alpha \in [-.5, .5) \end{cases} \tag{1}$$

where $round(\cdot)$ is the usual round operation, s_i has the closest index label to "β" and "α" is the value of the symbolic translation.

Proposition 1. *Let $S = \{s_0, ..., s_g\}$ be a linguistic term set and (s_i, α) be a 2-tuple. There is always a Δ^{-1} function, such that, from a 2-tuple it returns its equivalent numerical value $\beta \in [0, g] \subset \mathcal{R}$.*
Proof.
It is trivial, we consider the following function:

$$\Delta^{-1} : S \times [-.5, .5) \longrightarrow [0, g]$$

$$\Delta^{-1}(s_i, \alpha) = i + \alpha = \beta \tag{2}$$

Linguistic Computational Model Based on Linguistic 2-tuples

In this subsection, we present a computational technique to operate with the 2-tuples without loss of information. We shall present the following computations and operators:

1. Comparison of 2-tuples

The comparison of linguistic information represented by 2-tuples is carried out according to an ordinary lexicographic order.

Let (s_k, α_1) and (s_l, α_2) be two 2-tuples, with each one representing a counting of information:

- if $k < l$ then (s_k, α_1) is smaller than (s_l, α_2)
- if $k = l$ then
    1. if $\alpha_1 = \alpha_2$ then (s_k, α_1), (s_l, α_2) represents the same information
    2. if $\alpha_1 < \alpha_2$ then (s_k, α_1) is smaller than (s_l, α_2)
    3. if $\alpha_1 > \alpha_2$ then (s_k, α_1) is bigger than (s_l, α_2)

2. Negation operator of a 2-tuple

We define the negation operator over 2-tuples as:

$$Neg((s_i, \alpha)) = \Delta(g - (\Delta^{-1}(s_i, \alpha))) \qquad (3)$$

where $g + 1$ is the cardinality of S, $S = \{s_0, ..., s_g\}$.

3. Aggregation of 2-tuples

The aggregation of information consists of obtaining a value that summarizes a set of values, therefore, the result of the aggregation of a set of 2-tuples must be a 2-tuple. In the literature we can find many aggregation operators [29] which allow us to combine the information according to different criteria. The fuzzy linguistic representation model with 2-tuples has defined the functions Δ and Δ^{-1} that transform numerical values into 2-tuples and viceversa without loss of information, therefore any numerical aggregation operator can be easily extended for dealing with linguistic 2-tuples [12]. As example of linguistic 2-tuple aggregation operator we shall show the Linguistic Weighted Average operator.

Definition 3 [12]. *Let* $x = \{(r_1, \alpha_1), \ldots, (r_n, \alpha_n)\}$ *be a set of 2-tuples and* $W = \{(w_1, \alpha_1^w), ..., (w_n, \alpha_n^w)\}$ *be their linguistic 2-tuple associated weights. The 2-tuple linguistic weighted average* $\overline{x}_l^w$ *is:*

$$\overline{x}_l^w([(r_1, \alpha_1), (w_1, \alpha_1^w)]...[(r_n, \alpha_n), (w_n, \alpha_n^w)]) = \Delta\left(\frac{\sum_{t=1}^k \beta_i \cdot \beta_{W_i}}{\sum_{i=1}^n \beta_{W_i}}\right), \qquad (4)$$

with $\beta_i = \Delta^{-1}((r_i, \alpha_i))$ *and* $\beta_{W_i} = \Delta^{-1}(w_i, \alpha_i^w)$.

2.3 Linguistic Hierarchies

The *linguistic hierarchies* are a concept introduced in [4] for the design of *Hierarchical Systems of Linguistic Rules*. The hierarchical linguistic structure was also used in [14] to improve the precision in the processes of CW in linguistic multi-granular contexts.

A *linguistic hierarchy* is a set of levels, where each level is a linguistic term set with different granularity from the remaining of levels of the hierarchy. Each level belonging to a linguistic hierarchy is denoted as $l(t, n(t))$, being:

1. t, a number that indicates the level of the hierarchy,
2. $n(t)$, the granularity of the linguistic term set of the level t.

Here, we must point out that linguistic hierarchies deal with linguistic terms whose membership functions are triangular-shaped, symmetrical and uniformly distributed in $[0, 1]$. In addition, the linguistic term sets have an odd value of granularity representing the central label the value of *indifference*.

The levels belonging to a linguistic hierarchy are ordered according to their granularity, i.e., for two consecutive levels t and $t + 1$, $n(t + 1) > n(t)$. This provides a linguistic refinement of the previous level.

From the above concepts, we shall define a linguistic hierarchy, LH, as the union of all levels t:

$$LH = \bigcup_t l(t, n(t))$$

To build a linguistic hierarchy we must keep in mind that the hierarchical order is given by the increase of the granularity of the linguistic term sets in each level. Starting from a linguistic term set, S, over the universe of the discourse U in the level t:

$$S = \{s_0, ..., s_{n(t)-1}\}$$

being $s_k, (k = 0, ..., n(t) - 1)$ a linguistic term of S.

We extend the definition of S to a set of linguistic term sets, $S^{n(t)}$, each term set belongs to a level t of the hierarchy and has a granularity of uncertainty $n(t)$:

$$S^{n(t)} = \{s_0^{n(t)}, ..., s_{n(t)-1}^{n(t)}\}$$

The building of a linguistic hierarchy satisfies the following rules, that we call, *linguistic hierarchy basic rules*:

1. To preserve all *former modal points* of the membership functions of each linguistic term from one level to the following one.
2. To make *smooth transitions between successive levels*. The aim is to build a new linguistic term set, $S^{n(t+1)}$. A new linguistic term will be added between each pair of terms belonging to the term set of the previous level t. To carry out this insertion, we shall reduce the support of the linguistic labels in order to keep place for the new one located in the middle of them.

Generically, we can say that the linguistic term set of level $t+1$ is obtained from its predecessor t as:

$$l(t, n(t)) \rightarrow l(t+1, 2 \cdot n(t) - 1) \qquad (5)$$

Remark: Therefore we can say, that each label of the *level* t is a generating label of two labels in the next level, $t+1$ (excepting the central label).

Table 1 shows the granularity needed in each linguistic term set of the level t depending on the value $n(t)$ defined in the first level (3 and 7 respectively).

Table 1. Linguistic Hierarchies

	Level 1	Level 2	Level 3
$l(t, n(t))$	$l(1, 3)$	$l(2, 5)$	$l(3, 9)$
$l(t, n(t))$	$l(1, 7)$	$l(2, 13)$	

A graphical example of a linguistic hierarchy is shown in figure 2:

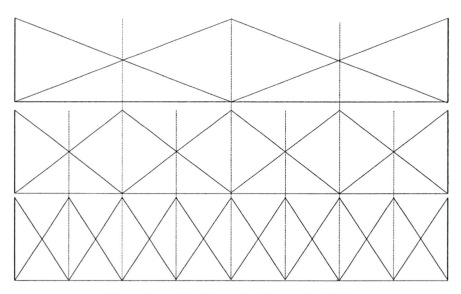

Fig. 2. Linguistic Hierarchy of 3, 5 and 9 labels

In [14] were defined transformation functions between labels from different levels to make processes of CW in multi-granular linguistic contexts without loss of information that will be useful in the intelligent agent model under multi-granular linguistic information.

Definition 4. Let $LH = \bigcup_t l(t, n(t))$ be a linguistic hierarchy whose linguistic term sets are denoted as $S^{n(t)} = \{s_0^{n(t)}, ..., s_{n(t)-1}^{n(t)}\}$, and let us consider the 2-tuple linguistic representation. The transformation function from a linguistic label in level t to a label in consecutive level $t+c$, with $c \in {-1, 1}$, is defined as:

$$TF_{t+c}^t : l(t, n(t)) \longrightarrow l(t + c, n(t + c)) \tag{6}$$

$$TF_{t+c}^t(s_i^{n(t)}, \alpha^{n(t)}) = \Delta\left(\frac{\Delta^{-1}(s_i^{n(t)}, \alpha^{n(t)}) \cdot (n(t + c) - 1)}{n(t) - 1}\right)$$

This transformation function was generalized to transform linguistic terms between any linguistic level in the linguistic hierarchy.

Definition 5. Let $LH = \bigcup_t l(t, n(t))$ be a linguistic hierarchy whose linguistic term sets are denoted as $S^{n(t)} = \{s_0^{n(t)}, ..., s_{n(t)-1}^{n(t)}\}$. The recursive transformation function between a linguistic label that belongs to level t and a label in level $t'=t+a$, with $a \in Z$, is defined as:

$$TF_{t'}^t : l(t, n(t)) \longrightarrow l(t', n(t'))$$

If $|a| > 1$ then

$$TF_{t'}^t(s_i^{n(t)}, \alpha^{n(t)}) = TF_{t'}^{t+\frac{t-t'}{|t-t'|}}(TF_{t+\frac{t-t'}{|t-t'|}}^t(s_i^{n(t)}, \alpha^{n(t)})) \tag{7}$$

If $|a| = 1$ then

$$TF_{t'}^t(s_i^{n(t)}, \alpha^{n(t)}) = TF_{t+\frac{t-t'}{|t-t'|}}^t(s_i^{n(t)}, \alpha^{n(t)})$$

This recursive transformation function can be easily defined in a non recursive way as follows:

$$TF_{t'}^t : l(t, n(t)) \longrightarrow l(t', n(t'))$$

$$TF_{t'}^t(s_i^{n(t)}, \alpha^{n(t)}) = \Delta\left(\frac{\Delta^{-1}(s_i^{n(t)}, \alpha^{n(t)}) \cdot (n(t') - 1)}{n(t) - 1}\right) \tag{8}$$

Proposition 2. *The transformation function between linguistic terms in different levels of the linguistic hierarchy is bijective:*

$$TF_t^{t'}(TF_{t'}^t(s_i^{n(t)}, \alpha^{n(t)})) = (s_i^{n(t)}, \alpha^{n(t)}) \tag{9}$$

This result guarantees the transformations between levels of a linguistic hierarchy are carried out without loss of information.

3 A Distributed Intelligent Agent Model for Information Gathering in Multi-Granular Linguistic Contexts

In this section, we present a linguistic agent model for gathering information on the Internet where the communication among the agents of different levels and between users and agents is carried out by using different label sets (multi-granular linguistic information) in order to allow a higher flexibility in the processes of communication of the system. We assume that in the agent system the importance degrees associated with the weighted user queries, the satisfaction degrees of weighted user queries and the relevance degree of the retrieved documents are expressed by means of linguistic values assessed in linguistic term sets with different granularity.

In the first subsection, the main notions of the concept *Intelligent agent* is set, in the second one an architecture for information gathering with these agents is proposed and finally, the process of information gathering that allows to manage the multi-granular linguistic communication in the distributed agent model is given.

3.1 Concept of Intelligent Software Agent

The intelligent software agents have been defined several times in the literature [22, 25, 28]. We are not to give a new definition of this concept, neither to review the ones given previously, but to set the main notions about those characteristics from every of these terms related to our specific purpose.

The concept of *agent* or rather *autonomous agent* must be the first one to be explained. This term, is strongly associated with the "behavior-based AI", as opposed to the "knowledge-based AI" [22], led by the expert systems. As Maes defines in [22], an agent is a system that tries to achieve some predefined goals in a complex and dynamic environment. Thus, depending on the environment, we can set the first big gap, by splitting the concept of agent in those called typically "robots", whose environment is basically physical, and those called "software agents", that inhabit in an environment consisting of computers and networks. Both concepts share one main characteristic: they are autonomous, i.e. they are able to operate and decide themselves the way to achieve their goals. However, as this feature is supposed to be inherent in an agent, an *autonomous agent* is usually called simply *agent*. As for the term *intelligent*, there are several discussions [25] about to consider whether an agent is intelligent by nature or not. We shall consider them as intelligent, since they present, in some sense, human behavior reducing the heaviest work of Internet users. Hence, the agents which with we are dealing with, are *intelligent agents*.

3.2 A Distributed Multi-Agent Architecture on the Internet

Most the designed intelligent agents nowadays are closely connected to the Internet. These agents do not only retrieve and filter information (in the sense of Web documents) [23], but also hand electronic mail, news lists, FAQ lists, ..., [19, 21, 28]. These are properly called *interface agents* [21], since they are more closely to the user. However, all the information that these agents get, come from somewhere or somewhat. There are servers through the Internet that proportionate these services of information, mail, news and FAQs. The agents closest to these data sources are called *information agents* [27]. Since Internet users can access to their interface or personal agents, as well as the general information agents, they feel completely lost and overloaded of information due to this avalanche of agents. This problem reveals the need of an organisation among the agents within Internet that implies both an agent hierarchy and architecture. Since the disposition of the elements taking part in the retrieval information process is distributed, it seems sensible to consider the architecture as a distributed one. Several architectures for these multi-agents distributed models have been proposed and reviewed [20, 24, 27, 28]. However, the architecture that fits better to our model is the one proposed by Sycara et al. in [27]. In this architecture, besides the aforementioned *interface* and *information agents*, the authors consider a third type of agents, the *task agents*. These agents deal with the decision-making process and the exchange of information with the information agents, resolving conflicts and fusing information, in order to release the interface agents of some tasks that make them ineffective.

A hierarchical model with five levels is proposed, as set out below:

- **Level 1:** *Internet Users*, which look for Web documents on the Internet by means of a weighted query where a set of terms related to the desired documents is specified.
- **Level 2:** *Interface Agents* (one for user, generally), that communicate the user's weighted query to the task agents, and filter the retrieved documents from task agents in order to give to the users those that satisfy better their needs.
- **Level 3:** *Task Agents* (one for interface agent, generally), that communicate the user's query to the information agents, and get those documents from every information agent that fulfills better the query, fusing them and resolving the possible conflicts among the information agents.
- **Level 4:** *Information Agents*, which receive the weighted query from the task agents, look for the information in the data sources, and give the retrieval documents back to the previous level.
- **Level 5:** *Information Sources*, consisting of all data sources within the Internet, such as databases and information repositories.

The scheme of this model can be observed in Figure 3.

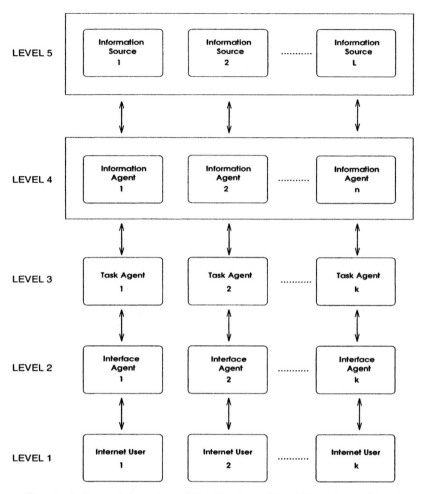

Fig. 3. A General Overview of the Distributed Intelligent Agent Model

3.3 Information Gathering By a Multi-Granular Linguistic Distributed Intelligent Agent Model

In the process of information gathering, as a response of a weighted user query on the presented agent model, there are two different parts:

- On the one hand, there is a communication between agents at levels 5-4 and 4-3, which is far from the user's participation, and where the question to be decided by the task agent is about which information agents would satisfy better the user's needs.
- On the other hand, there is a communication between agents at levels 3-2 and the user, where the information element is specifically the set of

retrieved documents that will be analyzed and filtered by the interface agents.

In [6, 7] we present some linguistic approaches to incorporate more flexibility in the communication carried out in the process of information gathering. The problem is that we always use the same linguistic domain to express the different assessments (importance degrees associated with the user queries, satisfaction degrees of user queries and relevance degrees of the retrieved documents) that appear in the communication process.

In this paper, we overcome the above problem by allowing that the different assessments of communication process can be assessed on different linguistic domains, i.e., by using multi-granular linguistic information. To do so, we propose that the system deals with a linguistic hierarchy $LH = \bigcup_t l(t, n(t))$, to express the different assessments, by using a level to assess each kind of assessment. For example, assuming the linguistic hierarchy shown in the Figure 2, the users can assess the importance degrees associated with the queries in the first level, the agents can assess the satisfaction degrees of a query in the second one and the relevance degrees of the retrieved documents in the third one.

Then, the description of the information gathering process related to a single user (see Figure 4) in a multi-granular linguistic context is as follows:

- **Step 1:** An *Internet user* makes a query to look for those documents related to the terms $\{t_1, t_2, ..., t_m\}$, which are weighted by a linguistic degree of importance $\{p_1, p_2, ..., p_m\}$, $p_i \in S^3$. Both set of values are given by the user to the *interface agent*.

- **Step 2:** The *interface agent* gives the terms and their importance weights to the *task agent*.

- **Step 3:** The *task agent* makes the query to all the information agents to which it is connected, and give them the terms $\{t_1, t_2, ..., t_m\}$.

- **Step 4:** All the *information agents* that have received the query, look for the information that better satisfies it in the information sources, and retrieve from them the documents.

- **Step 5:** The *task agent* receives from every *information agent* h a set of documents and their relevances (D^h, R^h) ordered decreasingly by relevance [26], where every document d_j^h has an associated linguistic degree of relevance $r_j^h \in S^9$ ($j = 1, ..., card(D^h)$) assessed in the set with maximum granularity of the linguistic hierarchy. It also receives a linguistic degree of satisfaction [2] $c_1^h, c_2^h, ..., c_m^h$, $c_i^h \in S^5$ (whose equivalent 2-tuples are $(c_1^h, 0), (c_2^h, 0), ..., (c_m^h, 0)$) of this set of documents with regard to every term of the query.

  - **Step 5.1:** The *task agent* aggregates both linguistic information weights, the satisfactions of the terms of the query from every *information agent*, $(c_i^h, \alpha), c_i^h \in S^5$, and the importance weights that the user assigned to these terms, $(p_i, \alpha), p_i \in S^3$, using the aggregation process for multi-granular linguistic information presented in [14]:

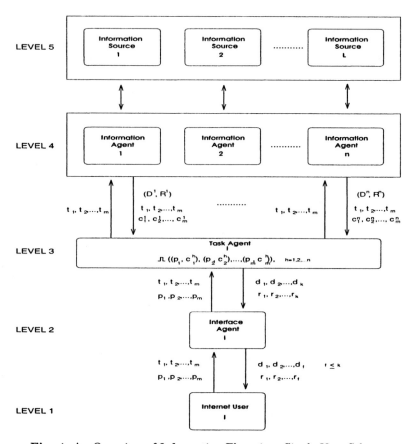

Fig. 4. An Overview of Information Flows in a Single User Scheme

1. *Normalization Phase*: the linguistic term set with highest granularity of the linguistic multi-granular context is chosen to make uniform the multi-granular linguistic information. Then, all the information are expressed in that linguistic term set by means of 2-tuples.

2. *Aggregation Phase*: through a 2-tuple aggregation operator the information is aggregated. In this paper we use the 2-tuple linguistic weighted average operator, $\bar{x}_l^w$, for combining the satisfactions of the terms of the query and the importance weights.

Let $\{[(p_1, \alpha), (c_1^h, \alpha)], \ ..., [(p_m, \alpha), (c_m^h, \alpha)]\}$, $p_i \in S^3$ and $c_i^h \in S^5$ be the set of pairs of linguistic 2-tuples of importance and satisfaction to be aggregated by the task agent for every information agent h. Then, for combining them first the linguistic values $(p_i, \alpha), p_i \in S^3$ and $(c_m^h, \alpha), c_m^h \in S^5$ are transformed in the linguistic term set with maximum granularity in the Linguistic Hierarchy, in this case S^9, ob-

taining their corresponding values $(\overline{p_i}, \alpha), \overline{p_i} \in S^9$ and $(\overline{c_m^h}, \alpha), \overline{c_m^h} \in S^9$.
Once the multi-granular information has been unified according to the
2-tuple linguistic weighted average operator definition, the aggregation
of the pair associated with every term is obtained as:

$$\lambda^h = \overline{x}_l^w([(\overline{p_1}, \alpha), (\overline{c_1^h}, \alpha)], \ ..., [(\overline{p_m}, \alpha), (\overline{c_m^h}, \alpha)]) \tag{10}$$

- **Step 5.2:** Once the *task agent* has calculated the overall performances
  $\{\lambda^1, \ ..., \lambda^n\}$, $\lambda^j \in S \times [-.5, .5)$ of the n *information agents* through
  the aggregation operator, it must decide which agent fulfil better the
  user's query. For this purpose, the task agent orders the performances
  decreasingly and obtains the vector $\{\Theta_1, \ ..., \Theta_n\}$, $\Theta_j \in S \times [-.5, .5)$ as
  follows:

$$\{\Theta_1, \ ..., \Theta_n\} = \sigma(\{\lambda^1, \ ..., \lambda^n\}) = \{\lambda^{\sigma(1)}, \ ..., \lambda^{\sigma(n)}\}, \tag{11}$$

where σ is a permutation over the set of linguistic 2-tuples $\{\lambda^1, \ ..., \lambda^n\}$
and

$$\lambda^{\sigma(j)} \leq \lambda^{\sigma(i)} \ \forall i \leq j.$$

In order to gather the better documents, the task agent may decide on
two alternatives.

- The first one is the selection of the information agent with the
  higher satisfaction of the query, Θ_1. This alternative presents a main
  drawback, as the set of documents of the selected agent contains
  some documents that, probably, will be less relevant to the query
  than some of the best documents of the rest of the information
  agents. This problem leads us to the second alternative, based on
  the selection of the best documents of every agent.
- In the second one, with the purpose of selecting a number of documents from every agent being proportional to the degree of satisfaction of such an agent:

$$P_s(\Theta_h) = \frac{\Delta^{-1}(\lambda^h)}{\sum_{i=1}^n \Delta^{-1}(\lambda^i)} \tag{12}$$

Finally, the number of documents, $k(D^h)$, that the *task agent* would
select from such an agent is expressed as:

$$k(D^h) = round(\frac{\sum_{i=1}^n card(D^i)}{n} \cdot P_s(\Theta_h)). \tag{13}$$

- **Step 6:** The *interface agent* receives from the *task agent* an ordered list
  of documents and their relevances $\{(d_j^h, r_j^h)\}$, where $d_j^h \in D^h$, $r_j^h \in R^h$,
  $1 \leq h \leq n$ and $j = 1, ..., k(D^h)$.
- **Step 7:** The *interface agent* filters these documents in order to give to the
  user only those documents that fulfill better his/her needs.

4 Example

In the following, an example of the application through this architecture is explained, using the Linguistic Hierarchy shown in Figure 2 whose terms in each level are:

- $S^3 = \{a_0 = Null\_Importance, a_1 = Medium\_Importance,$
  $a_2 = Total\_Importance\}$.

- $S^5 = \{b_0 = Null\_Satisfaction, b_1 = Low\_Satisfaction,$
  $b_2 = Medium\_Satisfaction, b_3 = High\_Satisfaction,$
  $b_4 = Total\_Satisfaction\}$.

- $S^9 = \{c_0 = Null\_Relevance, c_1 = Very\_Low\_Relevance, c_2 = Low\_Relevance,$
  $c_3 = More\_Less\_Low\_Relevance, c_4 = Medium\_Relevance,$
  $c_5 = More\_Less\_High\_Relevance, c_6 = High\_Relevance,$
  $c_7 = Very\_High\_Relevance, c_8 = Total\_Relevance\}$.

For this purpose, a view of a single user i will be considered, as it was set out in Figure 4. In this example, we will consider four information agents. Let us suppose a user making a query to Internet through an interface agent at the lowest levels of the presented architecture. The user may be interested in 'Agents', and more specifically, in 'Information Agents', to which the terms 'Agents' and 'Information' may be introduced as terms in the query. These terms may be weighted by means of linguistic 2-tuples related to importance assessed using the linguistic terms in S^3. Since the user is quite interested in the topic 'Agents' and, explicitly, in 'Information Agents', the labels associated to the query terms may be a_2 for the term 'Agents', and a_1 for the term 'Information'.

Therefore, the parameters which the user will communicate to the interface agent would be as follows:

$$(t_1, (p_1, \alpha)) = ('Agents', (a_2, 0)) \quad (t_2, (p_2, \alpha)) = ('Information', (a_1, 0))$$

The interface agent will go through the task agent, which will merely pass the terms of the query to the information agent level. The information agents search in the information source level those documents related to the terms of the query, and get a list with the most relevant links [2, 15]. For instance, each information agent h ($h = 1, \ldots, 4$) may retrieve a set of five links, D^h and their relevances R^h where each relevance degree r_j^h is assessed in S^9 (see Table 2).

Table 2. Sets of Documents for the Terms 'Agents' and 'Information'

(D^h, R^h)	d_j^h	r_j^h
(D^1, R^1)	http://phonebk.duke.edu/clients/tnfagent.html	c_6
	http://webhound.www.media.mit.edu/projects/webhound/doc/Webhound.html	c_6
	http://www.elet.polimi.it/section/compeng/air/agents/	c_5
	http://www.cs.bham.ac.uk/ámw/agents/links/	c_4
	http://groucho.gsfc.nasa.gov/Code_520/Code_522/Projects/Agents/	c_3
(D^2, R^2)	http://lcs.www.media.mit.edu/people/lieber/Lieberary/Letizia/Letizia.html	c_8
	http://www.osf.org/ri/contracts/6.Rationale.frame.html	c_7
	http://www.info.unicaen.fr/ serge/sma.html	c_7
	http://www.cs.umbc.edu/ cikm/1994/iia/papers/jain.html	c_3
	http://www.hinet.com/realty/edge/gallery.html	c_0
(D^3, R^3)	http://activist.gpl,ibm.com/WhitePaper/ptc2.htm	c_8
	http://www.cs.umbc.edu/ cikm/iia/submitted/viewing/chen.html	c_5
	http://www.psychology.nottingham.ac.uk:80/aigr/research/agents/agents.html	c_5
	http://netq.rowland.org/isab/isab.html	c_4
	http://maple.net/gbd/salagnts.html	c_3
(D^4, R^4)	http://www.ncsa.uiuc.edu/SDG/IT94/Proceedings/Agents/spetka/spetka.html	c_8
	http://mmm.wiwi.hu-berlin.de/MMM/cebit_engl.html	c_5
	http://foner.www.media.mit.edu/people/foner/Julia/subsection3_2_2.html	c_3
	http://www.cs.bham.ac.uk/ mw/agents/index.html	c_3
	http://www.ffly.com/html/About1.html	c_1

Each information agent h gives back to the task agent a set with the degree of relevance and the linguistic degree of satisfaction $c_i^h \in S^5$ of the set D^h with regard to every term p_i of the query, according to the following:

$$[(c_1^1, \alpha), (c_2^1, \alpha)] = [(b_2, 0), (b_1, 0)]$$
$$[(c_1^2, \alpha), (c_2^2, \alpha)] = [(b_3, 0), (b_3, 0)]$$
$$[(c_1^3, \alpha), (c_2^3, \alpha)] = [(b_3, 0), (b_2, 0)]$$
$$[(c_1^4, \alpha), (c_2^4, \alpha)] = [(b_3, 0), (b_1, 0)]$$

Once the task agent has received this information from the previous level, it aggregates both the satisfaction degrees and the importance degrees which had been obtained through the internet agent in an earlier step. To do so, it makes the information uniform in the term set with maximum granularity, S^9. Hence, the pairs of importance and satisfaction are aggregated by the task agent for every information agent h:

$$([[(\overline{p_1}, \alpha), (\overline{c_1^1}, \alpha)], [(\overline{p_2}, \alpha), (\overline{c_2^1}, \alpha)]]) = ([(c_8, 0), (c_4, 0)], [(c_4, 0), (c_2, 0)])$$
$$([[(\overline{p_1}, \alpha), (\overline{c_1^2}, \alpha)], [(\overline{p_2}, \alpha), (\overline{c_2^2}, \alpha)]]) = ([(c_8, 0), (c_6, 0)], [(c_4, 0), (c_6, 0)])$$
$$([[(\overline{p_1}, \alpha), (\overline{c_1^3}, \alpha)], [(\overline{p_2}, \alpha), (\overline{c_2^3}, \alpha)]]) = ([(c_8, 0), (c_6, 0)], [(c_4, 0), (c_4, 0)])$$
$$([[(\overline{p_1}, \alpha), (\overline{c_1^4}, \alpha)], [(\overline{p_2}, \alpha), (\overline{c_2^4}, \alpha)]]) = ([(c_8, 0), (c_6, 0)], [(c_4, 0), (c_2, 0)])$$

The aggregation of each pair is carried out through the 2-tuple linguistic weighted average, $\overline{x}_l^w$. Therefore, the overall fulfillment λ^h of the information agent h will be determined by the following expressions:

$$\lambda^1 = \overline{x}_l^w([(c_8,0),(c_4,0)],[(c_4,0),(c_2,0)]) = (c_3,.33)$$
$$\lambda^2 = \overline{x}_l^w([(c_8,0),(c_6,0)],[(c_4,0),(c_6,0)]) = (c_6,0)$$
$$\lambda^3 = \overline{x}_l^w([(c_8,0),(c_6,0)],[(c_4,0),(c_4,0)]) = (c_5,.33)$$
$$\lambda^4 = \overline{x}_l^w([(c_8,0),(c_6,0)],[(c_4,0),(c_2,0)]) = (c_5,-.33)$$

Hence, the overall performances of the information agents is:

$$\{\lambda^1,\lambda^2,\lambda^3,\lambda^4\} = \{(c_3,.33),(c_6,0),(c_5,.33),(c_5,-.33)\}$$

In the next step, the task agent would order these values decreasingly as follows:

$$\{\Theta_1,\Theta_2,\Theta_3,\Theta_4\} = \{\lambda^2,\lambda^3,\lambda^4,\lambda^1\} = \{(c_6,0),(c_5,.33),(c_5,-.33),(c_3,.33)\}$$

As it was explained in Section 3.3 (step 5.2), the task agent may decide on choosing the information agent with the highest performance, or select the best documents from all the agents, according to the performance of each one. In general, this last solution is most suitable when all the information agents present similar performances, as it is our case. Therefore, the task agent will calculate the probabilities of selection of the documents of each agent, according to the scheme of selection probabilities referenced in Step 5.2, which expression would set as follows:

$$P_s(\Theta_h) = \frac{\Delta^{-1}(\lambda^h)}{\sum_1^4 \Delta^{-1}(\lambda^i)},$$

Obtaining,

$$P_s(\Theta_1) = 0.137, P_s(\Theta_2) = 0.310, P_s(\Theta_3) = 0.276 \ and \ P_s(\Theta_4) = 0.241.$$

Finally, the task agent would calculate the number of documents $k(D^h), h = 1,\dots,n$ to select from each agent. The result of this computation would be:

$$k(D^1) = 1, k(D^2) = 2, k(D^3) = 1, k(D^4) = 1.$$

Hence, the final list of documents ordered by relevance that the interface agent would receive from the task agent would be:

$(d_1^2, r_1^2) = (http://lcs.www.media.mit.edu/people/lieber/Lieberary/Letizia/Letizia.html, c_8)$
$(d_1^3, r_1^3) = (http://www.activist.gpl.ibm.com/WhitePaper/ptc2.htm, c_8)$
$(d_1^4, r_1^4) = (http://www.ncsa.uiuc.edu/SDG/IT94/Proceedings/Agents/spetka/spetka.html, c_8)$
$(d_2^2, r_2^2) = (http://www.osf.org/ri/contracts/6.Rationale.frame.html, c_7)$
$(d_1^1, r_1^1) = (http://phonebk.duke.edu/clients/tnfagent.html, c_6)$

i.e., three documents with a total relevance degree, one document with an extremely high relevance degree and one document with a very high relevance degree.

In the last step of the information gathering process, the interface agent would filter this final ranked list of documents and would give to the user the most relevant documents.

This information gathering process guarantees that the user will receive the most relevant documents for his/her query, due to the fact, in step 5.2 we have chosen the second alternative proposed in the algorithm. Therefore, the ranking list of documents given to the user contains the documents with highest degree of satisfaction (to the query) according to all the agents avoiding a biassed selection of documents.

5 Concluding Remarks

We have presented a distributed intelligent agent system where the communication processes carried out in the information gathering are modelled by means of the multi-granular linguistic information. To do so, we have used the hierarchical linguistic contexts and the 2-tuple linguistic computational model.

We may stand out two main advantages of this proposal:

- The use of the multi-granular linguistic information allows a higher flexibility and expressiveness in the communication among the agents and between users and agents in the information gathering process.
- The use of the multi-granular linguistic information does not decrease the precision of system in its results.

References

1. P.P. Bonissone and K.S. Decker, Selecting Uncertainty Calculi and Granularity: An Experiment in Trading-off Precision and Complexity, in: L.H. Kanal and J.F. Lemmer, Eds., *Uncertainty in Artificial Intelligence* (North-Holland, 1986) 217-247.
2. G. Bordogna and G. Pasi, A Fuzzy Linguistic Approach Generalizing Boolean Information Retrieval: A Model and Its Evaluation, *J. of the American Society for Information Science* **44** (1993) 70-82.
3. W. Brenner, R. Zarnekow and H. Witting, *Intelligent Software Agent, Foundations and Applications.* Springer-Verlag. Berlin Heidelberg (1998)
4. O. Cordón, F. Herrera, and I. Zwir. Linguistic modeling by hierarchical systems of linguistic rules. *IEEE Transactions on Fuzzy Systems*, **10** (1) (2001) 2-20.
5. R. Degani and G. Bortolan, The Problem of Linguistic Approximation in Clinical Decision Making, *Int. J. of Approximate Reasoning* **2** (1988) 143-162 .
6. M. Delgado, F. Herrera, E. Herrera-Viedma, M.J. Martin-Bautista, M.A. Vila. Combining Linguistic Information in a Distributed Intelligent Agent Model for Information Gathering on the Internet, in: P.P. Wang, Ed., *Computing with Words,* (John Wiley & Son, 2001) 251-276.

7. M. Delgado, F. Herrera, E. Herrera-Viedma, M.J. Martin-Bautista, L. Mart-nez, M.A. Vila, A Communication Model Based on the 2-Tuple Fuzzy Linguistic Representation for a Distributed Intelligent Agent System on Internet, *Soft Computing* **6** (2002) 320-328.

8. B. Fazlollahi, R.M. Vahidov and R.A. Aliev, Multi-Agent Distributed Intelligent System Based on Fuzzy Decision Making, *Int. J. of Intelligent Systems* **15** (2000) 849-858.

9. J. Ferber, *Multi-Agent Systems: An Introduction to Distributed Artificial Intelligence.* Addison-Wesley Longman, New York (1999).

10. F. Herrera and E. Herrera-Viedma, Aggregation Operators for Linguistic Weighted Information, *IEEE Trans. on Systems, Man and Cybernetics, Part A: Systems* **27** (1997) 646-656.

11. F. Herrera, E. Herrera-Viedma and L. Martínez. A Fusion Approach for Managing Multi-Granularity Linguistic Term Sets in Decision Making, *Fuzzy Sets and Systems* **114** (2000) 43-58.

12. F. Herrera and L. Martínez, A 2-tuple Fuzzy Linguistic Representation Model for Computing with Words, *IEEE Transactions on Fuzzy Systems* **8:6** (2000) 746-752.

13. F. Herrera and L. Martínez, An Approach for Combining Linguistic and Numerical Information Based on 2-tuple Fuzzy Representation Model in Decision-Making, *Int. J. of Uncertainty, Fuzziness and Knowledge-Based Systems* **8:5** (2000) 539-562.

14. F. Herrera, L. Martínez. A model based on linguistic 2-tuples for dealing with multigranularity hierarchical linguistic contexts in Multiexpert Decision-Making. *IEEE Transactions on Systems, Man and Cybernetics. Part B: Cybernetics* **31:2** (2001) 227-234.

15. E. Herrera-Viedma, Modeling the Retrieval Process of an Information Retrieval System Using an Ordinal Fuzzy Linguistic Approach, *J. of the American Society for Information Science and Technology* **52:6** (2001) 460-475.

16. E. Herrera-Viedma and E. Peis, Evaluating the Informative Quality of Documents in SGML-Format Using Fuzzy Linguistic Techniques Based on Computing with Words, *Information Processing and Management* (2002), in press.

17. M. Kobayashi and K. Takeda, Information Retrieval on the Web *ACM Computing Surveys,* **32:2** (2000) 144-173.

18. S. Lawrence and C. Giles, Searching the Web: General and Scientific Information Access, *IEEE Communications Magazine* **37:1** (1998) 116-122.

19. H. Lieberman, Personal Assistants for the Web: A MIT perspective, in: M.Klusch, Ed., *Intelligent Information Agents* (Springer-Verlag, 1999) 279-292.

20. M. Lejter and T. Dean, A Framework for the Development of Multiagent Architectures, *IEEE Expert* (1996) 47-59.

21. P. Maes, Agents that Reduce Work and Information Overload, *Communications of the ACM* **37** (1994) 31-40.

22. P. Maes, Modeling Adaptive Autonomous Agents, in: C.G. Langton, Ed., *Artificial Life: An Overview* (MIT Press, 1995) 135-162.

23. M.J. Martin-Bautista, H.L. Larsen, M.A. Vila, A Fuzzy Genetic Algorithm to an Adaptive Information Retrieval Agent, *J. of the American Society for Information Science* **50**(9) (1999) 760-771.

24. A. Monkas, G. Zacharia, P. Maes, Amalthaea and Histos: Multiagent Systems for WWW Sites and Representation Recommendations, in: M.Klusch, Ed., *Intelligent Information Agents* (Springer-Verlag, 1999) 293-322.

25. C.J. Petrie, Agent-Based Engineering, the Web and Intelligence, *IEEE Expert* (1996) 24-29.

26. G. Salton and M.G. McGill, *Introduction to Modern Information Retrieval*, (McGraw-Hill, 1983).

27. K. Sycara, A. Pannu, M. Williamson and D. Zeng, Distributed Intelligent Agents, *IEEE Expert* (1996) 36-46.

28. M. Wooldridge and N. Jennings, Intelligent Agents: Theory and Practice, *The Knowledge Engineering Review* **10** (1995) 115-152.

29. R.R. Yager, On Ordered Weighted Averaging Aggregation Operators in Multi-criteria Decision Making, *IEEE Transactions on Systems, Man, and Cybernetics* **18** (1988) 183-190.

30. R.R. Yager, An Approach to Ordinal Decision Making, *Int. J. of Approximate Reasoning* **12** (1995) 237-261.

31. R.R. Yager, Protocol for Negotiations Among Multiple Intelligent Agents, in: J. Kacprzyk, H. Nurmi and M. Fedrizzi, Eds., *Consensus Under Fuzziness* (Kluwer Academic Publishers, 1996) 165-174.

32. R.R. Yager, Intelligent Agents for World Wide Web Advertising Decisions, *International J. of Intelligent Systems* **12** (1997) 379-390.

33. R.R. Yager, Fusion of Multi-Agent Preference Orderings, *Fuzzy Sets and Systems* **112** (2001) 1-12.

34. L.A. Zadeh, The Concept of a Linguistic Variable and Its Applications to Approximate Reasoning. Part I, *Information Sciences* **8** (1975) 199-249, Part II, *Information Sciences* **8** (1975) 301-357, Part III, *Information Sciences* **9** (1975) 43-80.

A Time-Completeness Tradeoff on Fuzzy Web-Browsing Mining

Tzung-Pei Hong[1], Kuei-Ying Lin[2], and Shyue-Liang Wang[3]

[1] Department of Electrical Engineering, National University of Kaohsiung
Kaohsiung, 811, Taiwan, R.O.C. tphong@nuk.edu.tw
[2] Telecommunication Laboratories, Chunghwa Telecom Ltd. Chung-li, 326,
Taiwan, R.O.C. ying120@cht.com.tw
[3] Department of Information Management, National University of Kaohsiung
Kaohsiung, 811, Taiwan, R.O.C. slwang@nuk.edu.tw

Abstract. World-wide-web applications have grown very rapidly and have made a significant impact on computer systems. Among them, web browsing for useful information may be most commonly seen. Due to its tremendous amounts of use, efficient and effective web retrieval has thus become a very important research topic in this field. In the past, we proposed a web-mining algorithm for extracting interesting browsing patterns from log data in web servers. It integrated fuzzy-set concepts and data mining approach to achieve this purpose. In that algorithm, each web page used only the linguistic term with the maximum cardinality in the mining process. The number of items was thus the same as that of the original web page, making the processing time reduced. The fuzzy browsing patterns derived in this way are, however, not complete, meaning some possible patterns may be missed. This paper thus modifies it and proposes a new fuzzy web-mining algorithm for extracting all possible fuzzy interesting knowledge from log data in web servers. The proposed algorithm can derive a more complete set of browsing patterns but with more computation time than the previous method. Trade-off thus exists between the computation time and the completeness of browsing patterns. Choosing an appropriate mining method thus depends on the requirements of the application domains.

Keywords: web mining, fuzzy set, browsing pattern, web page, world wide web.

1 Introduction

Techniques of web mining have recently been requested and developed for efficient and effective web retrieval. Cooley et. al. divided web mining into two classes: web-content mining and web-usage mining. Web-content mining focuses on information discovery from sources across the World Wide Web. On the other hand, web-usage mining emphasizes on the automatic discovery of user access patterns from web servers [11]. Many web-mining approaches for finding sequential patterns and user interesting information from the World Wide Web have also been proposed [5, 6, 8, 9, 10].

Fuzzy set theory is being used more and more frequently in intelligent systems because of its simplicity and similarity to human reasoning [22]. The theory has been applied in fields such as manufacturing, engineering, control, diagnosis, economics, among others [14, 22, 23, 31, 32]. Several fuzzy learning algorithms for inducing rules from a given set of data have been designed and used for specific domains [2, 3, 4, 12, 13, 15, 16, 17, 20, 21, 26, 28]. Strategies based on decision trees were proposed in [7, 8, 24, 25, 26, 30, 31]. Wang et al. proposed a fuzzy version space learning strategy for managing vague information [26]. Hong et al. also proposed a fuzzy mining approach for finding association rules from transactions [18].

In the past, we proposed a sophisticated fuzzy web-mining algorithm able to deal with data logs on web servers [19]. The browsing time of a customer on each web page is used to analyze the retrieval behavior of a web site. Since the data collected are numeric, fuzzy concepts are used to process them and to form linguistic terms. In that algorithm, each web page used only the linguistic term with the maximum cardinality in the mining process. The number of items was thus the same as that of the original web pages, making the processing time reduced. The fuzzy browsing patterns derived in that way are however not complete, meaning some possible fuzzy browsing patterns may be missed. This paper thus modifies it and proposes a new fuzzy web-mining algorithm for extracting all possible fuzzy interesting patterns from log data on web servers. The proposed algorithm considers all the important linguistic terms in the mining process, thus being able to derive a more complete set of browsing patterns but with more computation time than the previous method. Trade-off thus exists between the computation time and the completeness of browsing patterns.

The remaining parts of this paper are organized as follows. Agrawal and Srikant's method for mining sequential patterns is introduced in Section 2. Fuzzy-set concepts are reviewed in Section 3. Notation used in this paper is defined in Section 4. An algorithm for mining a complete set of fuzzy browsing patterns from log data on servers is proposed in Section 5. An example to illustrate the proposed algorithm is given in Section 6. Experimental results for showing the time-completeness trade-off effect are described in Section 7. Discussion and Conclusion are stated in Section 8.

2 Review of Agrawal and Srikant's Mining Approach

Agrawal and Srikant proposed a mining algorithm to discover sequential patterns from a set of transactions [1]. Five phases are included in their approach.

In the first phase, the transactions are sorted first by customer ID as the major key and then by transaction time as the minor key. This phase thus converts the original transactions into customer sequences. In the second phase, the set of all large itemsets are found from the customer sequences by comparing their counts with a predefined support parameter α. This phase is similar to the process of mining association rules. Note that when an itemset occurs more than one time in a customer sequence, it is counted once for this customer sequence. In the third phase, each large itemset is mapped to a contiguous integer and the original customer sequences are transformed into the mapped integer sequences. In the fourth phase, the set of transformed integer sequences are used to find large sequences among them. In the fifth phase, the maximally large sequences are then derived and output to users.

3 Review of Fuzzy Set Concepts

Fuzzy set theory was first proposed by Zadeh in 1965 [32]. Fuzzy set theory is primarily concerned with quantifying and reasoning using natural language in which words can have ambiguous meanings. This can be thought of as an extension of traditional crisp sets, in which each element must either be in or not be in a set.

Formally, the process by which individuals from a universal set X are determined to be either members or non-members of a crisp set can be defined by a characteristic or discrimination function [32]. For a given crisp set A, this function assigns a value $\mu_A(x)$ to every $x \in X$ such that

$$\mu_A(x) = \begin{cases} 1 \text{ if and only if } x \in A, \\ 0 \text{ if and only if } x \notin A. \end{cases} \tag{1}$$

Thus, the function maps elements of the universal set to the set containing 0 and 1. This function can be generalized such that the values assigned to the elements of the universal set fall within specified ranges, referred to as the membership grades of these elements in the set. Larger values denote higher degrees of set membership. Such a function is called the membership function, $\mu_A(x)$, by which a fuzzy set A is usually defined. This function is thus represented by

$$\mu_A : X \rightarrow [0,1], \tag{2}$$

where [0, 1] denotes the interval of real numbers from 0 to 1, inclusive. The function can also be generalized to any real interval instead of [0, 1].

A special notation is often used in the literature to represent fuzzy sets. Assume that x_1 to x_n are the elements in fuzzy set A, and μ_1 to μ_n are, respectively, their grades of membership in A. A is then represented as follows:

$$A = \mu_1/x_1 + \mu_2/x_2 + \cdots + \mu_n/x_n. \tag{3}$$

An $\alpha - cut$ of a fuzzy set A is a crisp set A_α that contains all elements in the universal set X with membership grades in A greater than or equal to a specified value α. This definition can be written as:

$$A_\alpha = \{x \in X | \mu_A(x) \geq \alpha\}. \tag{4}$$

The *scalar cardinality* of a fuzzy set A defined on a finite universal set X is the summation of the membership grades of all the elements of X in A. Thus,

$$|A| = \sum_{x \in X} \mu_A(x). \tag{5}$$

Three operations commonly used on fuzzy sets are *complementation*, *union* and *intersection*, as proposed by Zadeh.

1. The complementation of a fuzzy set A is denoted by $\neg A$, and the membership function of $\neg A$ is given by:

$$\mu_{\neg A}(x) = 1 - \mu_A(x), \forall x \in X. \tag{6}$$

2. The intersection of two fuzzy sets A and B is denoted by $A \cap B$, and the membership function of $A \cap B$ is given by:

$$\mu_{A \cap B}(x) = min\{\mu_A(x), \mu_B(x)\}, \forall x \in X. \tag{7}$$

3. The union of fuzzy sets A and B is denoted by $A \cup B$, and the membership function of $A \cup B$ is given by:

$$\mu_{A \cup B}(x) = max\{\mu_A(x), \mu_B(x)\}, \forall x \in X. \tag{8}$$

The above concepts will be used in our proposed algorithm to mine clients' linguistic browsing behavior on a web site.

4 Notation

The notation used in this paper is defined below.

n: *the total number of log data;*
m: *the total number of web pages in the log data;*
c: *the total number of clients in the log data;*
n_i: *the number of log data from the i-th client, $1 \leq i \leq c$;*
D_i: *the browsing sequence of the i-th client, $1 \leq i \leq c$;*
D_{id}: *the d-th log transaction in D_i, $1 \leq d \leq n_i$;*
I^g : *the g-th web page, $1 \leq g \leq m$;*
R^{gk}: *the k-th fuzzy region for I^g, $1 \leq k \leq |I^g|$, where $|I^g|$ is the number of fuzzy regions for I^g;*

v_{id}^g: the browsing duration of web page I^g in D_{id};
f_{id}^g: the fuzzy set converted from v_{id}^g;
$f_{id}^{g_k}$: the membership value of v_{id}^g in region R^{g_k};
$f_i^{g_k}$: the membership value of region R^{g_k} in the i-th client sequence D_i;
$count^{g_k}$: the scalar cardinality of region R^{g_k};
α: the predefined minimum support value;
C_r: the set of candidate sequences with r items;
L_r: the set of large sequences with r items.

5 Fuzzy Web Mining for Browsing Patterns

The web-mining algorithm proposed in [19] used only the linguistic term with the maximum count for each web page in the mining process. In this paper, all the linguistic terms are used. The computation is thus more complex than that in [19] since all possible linguistic terms are used in calculating large itemsets, but the derived set of linguistic browsing patterns is more complete. The details of the proposed web-mining algorithm are described below.

The web-mining algorithm for all possible linguistic browsing patterns:

INPUT: A server log, a set of membership functions, a predefined minimum support value α.

OUTPUT: set of all possible linguistic browsing patterns.

STEP 1: Select the transactions with file names including .asp, .htm, .html, .jva, .cgi, .php and closing connection from the log data; keep only the fields *date*, *time*, *client-ip* and *file-name*; denote the resulting log data as D.

STEP 2: Transform the *client-ips* into contiguous integers (called encoded client ID) for convenience, according to their first browsing time. Note that the same *client-ip* with two closing connections is given two integers.

STEP 3: Sort the resulting log data first by encoded client ID and then by date and time.

STEP 4: Calculate the time durations of the web pages browsed by each encoded client ID from the time interval between a web page and its next page.

STEP 5: Form a browsing sequence D_j for each client c_j by sequentially listing his/her n_j tuples (web page, duration), where n_j is the number of web pages browsed by client c_j. Denote the d-th tuple in D_j as D_{jd}.

STEP 6: Transform the time duration v_{id}^g of the file name I^g appearing in D_{id} into a fuzzy set f_{id}^g represented as

$$\left(\frac{f_{id}^{g_1}}{R^{g_1}} + \frac{f_{id}^{g_2}}{R^{g_2}} + \cdots + \frac{f_{id}^{g_i}}{R^{g_i}} \right) \tag{9}$$

using the given membership functions, where I_g is the g-th file name, R^{g_k} is the k-th fuzzy region of item I_g, $f_{id}^{g_k}$ is v_{id}^g 's fuzzy membership value in region R^{g_k}, and l is the number of fuzzy regions for I^g.

STEP 7: Find the membership value $f_i^{g_k}$ of each region R^{g_k} in each browsing sequence D_i as

$$f_i^{g_k} = Max_{d=1}^{|D_i|} f_{id}^{g_k} \tag{10}$$

where $|Di|$ is the number of tuples in D_i.

STEP 8: Calculate the scalar cardinality of each region R^{g_k} as

$$count^{g_k} = \sum_{i=1}^{c} f_i^{g_k} \tag{11}$$

where c is the number of browsing sequences.

STEP 9: Check whether the value $count^{g_k}$ of each fuzzy region R^{g_k} is larger than or equal to the predefined minimum support value α. If a region R^{g_k} is equal to or greater than α, put it in the set of large 1-sequences (L_1). That is,

$$L_1 = \{R^{g_k} | count^{g_k} \geq \alpha, 1 \leq g \leq m, 1 \leq k \leq l\}. \tag{12}$$

STEP 10: If L_1 is null, then exit the algorithm; otherwise, do the next step.

STEP 11: Set $r = 1$, where r is used to represent the length of sequential patterns currently kept.

STEP 12: Generate the candidate set C_{r+1} from L_r in a way similar to that in the *aprioriall* algorithm [4]. Restated, the algorithm first joins L_r and L_r, under the condition that $r-1$ items in the two itemsets are the same and with the same orders. Different permutations represent different candidates. The algorithm then keeps in C_{r+1} the sequences which have all their subsequences of length r existing in L_r.

STEP 13: Do the following substeps for each newly formed $(r+1)$-sequence s with contents $(s_1, s_2, \cdots, s_{r+1})$ in C_{r+1}:

(a) Calculate the fuzzy value f_i^s of s in each browsing sequence D_i as

$$f_i^s = Min_{k=1}^{r+1} f_i^{s_k}, \tag{13}$$

where region s_k must appear after region s_{k-1} in D_i. If two or more same subsequences exist in D_i, then f_i^s is the maximum fuzzy value among those of these subsequences.

(b) Calculate the scalar cardinality of s as:

$$count^s = \sum_{i=1}^{c} f_i^s, \qquad (14)$$

where c is number of browsing sequences.

(c) If $count^s$ is larger than or equal to the predefined minimum support value α, put s in L_{r+1}.

STEP 14: IF L_{r+1} is null, then do the next step; otherwise, set $r = r + 1$ and repeat STEPs 12 to 14.

STEP 15: Output the maximally large q-sequences, $q \geq 2$, to web-site mangers as browsing patterns.

After STEP 15, the sequential browsing patterns output can serve as meta-knowledge concerning the given transactions.

6 An Example

In this section, a simple example is given to show how the proposed algorithm can be used to generate sequential patterns for clients' browsing behavior according to the log data in a web server. A part of the log data is shown in Table 1.

Each transaction in the log data includes fields *date*, *time*, *client-ip*, *server-ip*, *server-port* and *file-name*, among others. Only one file name is contained in each transaction. For example, the user in the client-ip 140.127.194.82 browsed the file *cheap.htm* at 05:39:54 on March 1st, 2001. Assume the fuzzy membership functions for a browsing duration on a web page are shown in Figure 1.

In Figure 1, the browsing duration is divided into three fuzzy regions: *Short*, *Middle* and *Long*. Thus, three fuzzy membership values are produced for each duration according to the predefined membership functions. For the log data shown in Table 1, the proposed web mining algorithm proceeds as follows.

STEP 1: The transactions with file names .asp, .htm, .html, .jva, .cgi, .php and closing connection are selected for mining. Only the four fields *date*, *time*, *cilent-ip* and *file-name* are kept. Assume the resulting log data from Table 1 are shown in Table 2.

STEP 2: The values of field *client-ip* are transformed into contiguous integers according to each client's first browsing time. Results for Table 2 are

Table 1. A part of the log data used in the example

Date	Time	Client-IP	Server-IP	Port	File-name	
2001-03-01	05:39:54	140.127.194.82	140.127.194.88	80	cheap.htm	$\cdots$
2001-03-01	05:39:55	140.127.194.82	140.127.194.88	80	home-bg1.jpg	$\cdots$
2001-03-01	05:39:56	140.127.194.127	140.127.194.88	80	inside.htm	$\cdots$
$\cdots$	$\cdots$	$\cdots$	$\cdots$	$\cdots$	$\cdots$	$\cdots$
2001-03-01	05:40:26	140.127.194.127	140.127.194.88	80	person.asp	$\cdots$
$\cdots$	$\cdots$	$\cdots$	$\cdots$	$\cdots$	$\cdots$	$\cdots$
2001-03-01	05:40:53	140.127.194.82	140.127.194.88	80	line1.gif	$\cdots$
$\cdots$	$\cdots$	$\cdots$	$\cdots$	$\cdots$	$\cdots$	$\cdots$
2001-03-01	05:41:08	140.127.194.128	140.127.194.88	80	cheap.htm	$\cdots$
$\cdots$	$\cdots$	$\cdots$	$\cdots$	$\cdots$	$\cdots$	$\cdots$
2001-03-01	05:48:13	140.127.194.22	140.127.194.88	80	cheap.htm	$\cdots$
$\cdots$	$\cdots$	$\cdots$	$\cdots$	$\cdots$	$\cdots$	$\cdots$
2001-03-01	05:48:38	140.127.194.44	140.127.194.88	80	closing connection	$\cdots$
$\cdots$	$\cdots$	$\cdots$	$\cdots$	$\cdots$	$\cdots$	$\cdots$
2001-03-01	05:50:13	140.127.194.20	140.127.194.88	80	search.asp	$\cdots$
$\cdots$	$\cdots$	$\cdots$	$\cdots$	$\cdots$	$\cdots$	$\cdots$
2001-03-01	05:53:13	140.127.194.20	140.127.194.88	80	closing connection	$\cdots$

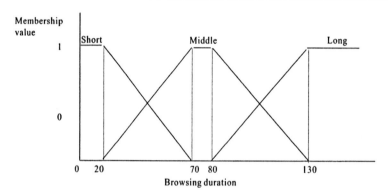

Fig. 1. The membership functions used in this example

shown in Table 3. Six clients logged in the web server and five web pages including *homepage.htm, longin.htm, search.asp, cheap.htm* and *person.asp* were browsed in this example.

STEP 3: The resulting log data in Table 3 are then sorted first by encoded client ID and then by date and time. Results are shown in Table 4.

STEP 4: The time durations of the web pages browsed by each encoded client ID are calculated. Take the first web page browsed by client 1 as an example. Client 1 retrieves the file *cheap.htm* at 05:39:54 on March 1st, 2001 and the next file *inside.htm* at 05:41:54 on March 1st, 2001. The duration of

Table 2. The resulting log data for web mining.

Date	Time	Client-ip	File-name
2001-03-01	05:39:54	140.127.194.182	cheap.htm
2001-03-01	05:39:56	140.127.194.128	inside.htm
2001-03-01	05:40:26	140.127.194.128	person.asp
2001-03-01	05:41:08	140.127.194.128	cheap.htm
2001-03-01	05:41:30	140.127.194.22	homepage.htm
2001-03-01	05:41:54	140.127.194.82	inside.htm
2001-03-01	05:42:25	140.127.194.82	cheap.htm
2001-03-01	05:42:46	140.127.194.128	search asp
2001-03-01	05:43:02	140.127.194.22	cheap.htm
2001-03-01	05:43:46	140.127.194.44	inside.htm
2001-03-01	05:44:06	140.127.194.44	search asp
2001-03-01	05:44:07	140.127.194.82	closing connection
2001-03-01	05:44:17	140.127.194.128	closing connection
2001-03-01	05:44:31	140.127.194.22	search.asp
2001-03-01	05:45:21	140.127.194.22	closing connection
2001-03-01	05:45:47	140.127.194.44	person.asp
2001-03-01	05:46:13	140.127.194.38	cheap.htm
2001-03-01	05:47:45	140.127.194.44	inside.htm
2001-03-01	05:47:53	140.127.194.38	inside.htm
2001-03-01	05:47:56	140.127.194.44	search.asp
2001-03-01	05:48:13	140.127.194.20	cheap.htm
2001-03-01	05:48:19	140.127.194.38	search.asp
2001-03-01	05:48:38	140.127.194.44	closing connection
2001-03-01	05:49:33	140.127.194.38	closing connection
2001-03-01	05:50:13	140.127.194.20	search.asp
2001-03-01	05:51:14	140.127.194.20	person.asp
2001-03-01	05:53:16	140.127.194.20	inside.htm
2001-03-01	05:53:33	140.127.194.20	closing connection

cheap.htm for client 1 is then 120 seconds (2001/03/01, 05:39:54 - 2001/03/01, 05:41:54).

Simple symbols are used here to represent web pages for convenience. Let A, B, C, D and E respectively represent *homepage.htm*, *inside.htm*, *search.asp*, *cheap.htm* and *person.asp*. The durations of all pages browsed by each client ID are shown in Table 5.

STEP 5: The web pages browsed by each client are listed as a browsing sequence. Each tuple is represented as (web page, duration). The resulting browsing sequences from Table 5 are shown in Table 6.

STEP 6: The time durations of the file names in each browsing sequence are represented as fuzzy sets. Take the web page D in the first browsing sequence as an example. The time duration "120" of file D is converted into the fuzzy set $\left(\frac{0.0}{Short} + \frac{0.2}{Middle} + \frac{0.8}{Long} \right)$ by the given membership functions (Figure 1).

Table 3. Transforming the values of field *client-ip* into contiguous integers

Date	Time	Client ID	File-name
2001-03-01	05:39:54	1	cheap.htm
2001-03-01	05:39:56	2	inside.htm
2001-03-01	05:40:26	2	person.asp
2001-03-01	05:41:08	2	cheap.htm
2001-03-01	05:41:30	3	homepage.htm
2001-03-01	05:41:54	1	inside.htm
2001-03-01	05:42:25	1	cheap.htm
2001-03-01	05:42:44	2	search asp
2001-03-01	05:43:02	3	cheap.htm
2001-03-01	05:43:46	4	inside.htm
2001-03-01	05:44:06	4	search asp
2001-03-01	05:44:07	1	closing connection
2001-03-01	05:44:17	2	closing connection
2001-03-01	05:44:31	3	search.asp
2001-03-01	05:45:21	3	closing connection
2001-03-01	05:45:47	4	person.asp
2001-03-01	05:46:13	5	cheap.htm
2001-03-01	05:47:45	4	inside.htm
2001-03-01	05:47:50	5	inside.htm
2001-03-01	05:47:56	4	search.asp
2001-03-01	05:48:13	6	cheap.htm
2001-03-01	05:48:19	5	search.asp
2001-03-01	05:48:38	4	closing connection
2001-03-01	05:49:33	5	closing connection
2001-03-01	05:50:13	6	search.asp
2001-03-01	05:51:14	6	person.asp
2001-03-01	05:53:16	6	inside.htm
2001-03-01	05:53:33	6	closing connection

This step is repeated for the other files and browsing sequences. The results are shown in Table 7.

STEP 7: The membership value of each region in each browsing sequence is found. Take the region $D.Middle$ for client 1 as an example. Its membership value is max(0.2, 0.0, 0.6)=0.6. The membership values of the other regions can be similarly calculated.

STEP 8: The scalar cardinality of each fuzzy region in all the browsing sequences is calculated as the *count* value. Take the fuzzy region $D.Middle$ as an example. Its scalar cardinality $= (0.6 + 0.6 + 0.8 + 0.0 + 0.6 + 0.2) = 2.8$. This step is repeated for the other regions, and the results are shown in Table 8.

STEP 9: The count of each region in Table 8 is checked against the predefined minimum support value α. Assume α is set at 2 in this example. Since

Table 4. Browsed pages sorted first by client ID and then by data and time

Date	Time	Client ID	File-name
2001-03-01	05:39:54	1	cheap.htm
2001-03-01	05:41:54	1	inside.htm
2001-03-01	05:42:25	1	cheap.htm
2001-03-01	05:44:07	1	closing connection
2001-03-01	05:39:56	2	inside.htm
2001-03-01	05:40:26	2	person.asp
2001-03-01	05:41:08	2	cheap.htm
2001-03-01	05:42:46	2	search asp
2001-03-01	05:44:17	2	closing connection
2001-03-01	05:41:30	3	homepage.htm
2001-03-01	05:43:02	3	cheap.htm
2001-03-01	05:44:31	3	search.asp
2001-03-01	05:54:21	3	closing connection
2001-03-01	05:43:46	4	inside.htm
2001-03-01	05:44:06	4	search asp
2001-03-01	05:45:47	4	person.asp
2001-03-01	05:47:45	4	inside.htm
2001-03-01	05:47:56	4	search.asp
2001-03-01	05:48:38	4	closing connection
2001-03-01	05:46:13	5	cheap.htm
2001-03-01	05:47:53	5	inside.htm
2001-03-01	05:48:19	5	search.asp
2001-03-01	05:49:33	5	closing connection
2001-03-01	05:48:13	6	cheap.htm
2001-03-01	05:50:13	6	search.asp
2001-03-01	05:51:14	6	person.asp
2001-03-01	05:53:16	6	inside.htm
2001-03-01	05:53:33	6	closing connection

the count values of *B.Short*, *C.Middle*, *D.Middle* and *D.Long* are larger than 2, these regions are put in L_1.

STEP 10: Since L_1 is not null, the next step is executed.

STEP 11: Set $r = 1$, where r is used to represent the length of sequential patterns currently kept.

STEP 12: The candidate set C_2 is generated from L_1 as follows:

(*B.Short*, *B.Short*), (*B.Short*, *C.Middle*), (*C.Middle*, *B.Short*), $\cdots$, (*D.Long*, *D.Middle*), (*D.Long*, *D.Long*).

STEP 13: The following substeps are done for each newly formed candidate 2-sequence in C_2:

(a) The fuzzy membership value of each candidate 2-sequence in each browsing sequence is calculated. Here, the minimum operator is used for the intersection. Take the sequence (*B.Short*, *C.Middle*) as an exam-

127

Table 5. The web pages browsed with their durations.

Client ID	(Web page, Duration)
1	(D, 120)
1	(B, 31)
1	(D, 102)
2	(B, 30)
2	(E, 42)
2	(D, 98)
2	(C, 91)
3	(A, 92)
3	(D, 89)
3	(C, 50)
4	(B, 20)
4	(C, 101)
4	(E, 118)
4	(B, 11)
4	(C, 42)
5	(D, 100)
5	(B, 29)
5	(C, 74)
6	(D, 120)
6	(C, 61)
6	(E, 122)
6	(B, 17)

Table 6. The browsing sequences formed from Table 5.

Client ID	Browsing Sequence
1	(D, 120) (B, 31) (D, 102)
2	(B, 30) (E, 42) (D, 98) (C, 91)
3	(A, 92) (D, 89) (C, 50)
4	(B, 20) (C, 101) (E, 118) (B, 11) (C, 42)
5	(D, 100) (B, 29) (C, 74)
6	(D, 120) (C, 61) (E, 122) (B, 17)

ple. Its membership value in the fourth browsing sequence is calculated as: $max[min(1.0, 0.6), min(1.0, 0.4), min(1.0, 0.4)] = 0.6$ since there are three subsequences of $(B.Short, C.Middle)$ in that browsing sequence. The results for sequence $(B.Short, C.Middle)$ in all the browsing sequences are shown in Table 9.

(b) The scalar cardinality (count) of each candidate 2-sequence in C_2 is calculated. Results for this example are shown in Table 10.

(c) Since only the counts of the 2-sequences $(B.Short, C.Middle)$, $(D.Middle, C.Middle)$, $(D.Long, B.Short)$ and $(D.Long, C.Middle)$ are larger than the predefined minimum support value 2, they are thus kept in L_2.

Table 7. The fuzzy sets transformed from the browsing sequences.

Client ID	Fuzzy Sets
1	$\left(\frac{0.2}{D.Middle} + \frac{0.8}{D.Long}\right)\left(\frac{0.8}{B.Short} + \frac{0.2}{B.Middle}\right)\left(\frac{0.6}{D.Middle} + \frac{0.4}{D.Long}\right)$
2	$\left(\frac{0.8}{B.Short} + \frac{0.2}{B.Middle}\right)\left(\frac{0.6}{E.Short} + \frac{0.4}{E.Middle}\right)\left(\frac{0.6}{D.Middle} + \frac{0.4}{D.Long}\right)$ $\left(\frac{0.8}{C.Middle} + \frac{0.2}{C.Long}\right)$
3	$\left(\frac{0.8}{A.Middle} + \frac{0.2}{A.Long}\right)\left(\frac{0.8}{D.Middle} + \frac{0.2}{D.Long}\right)\left(\frac{0.4}{C.Short} + \frac{0.6}{C.Middle}\right)$
4	$\left(\frac{1.0}{B.Short}\right)\left(\frac{0.6}{C.Middle} + \frac{0.4}{C.Long}\right)\left(\frac{0.2}{E.Middle} + \frac{0.8}{E.Long}\right)$ $\left(\frac{1.0}{B.Short}\right)\left(\frac{0.6}{C.Short} + \frac{0.4}{C.Middle}\right)$
5	$\left(\frac{0.6}{D.Middle} + \frac{0.4}{D.Long}\right)\left(\frac{0.8}{B.Short} + \frac{0.2}{B.Middle}\right)\left(\frac{1.0}{C.Middle}\right)$
6	$\left(\frac{0.2}{D.Middle} + \frac{0.8}{D.Long}\right)\left(\frac{0.2}{C.Short} + \frac{0.8}{C.Middle}\right)$ $\left(\frac{0.2}{E.Middle} + \frac{0.8}{E.Long}\right)\left(\frac{1.0}{B.Short}\right)$

Table 8. The counts of the fuzzy regions

Region	Count	Region	Count	Region	Count
A.Short	0.0	C.Short	1.2	E.Short	0.6
A.Midlle	0.8	C.Midlle	3.8	E.Midlle	0.8
A.Long	0.2	C.Long	0.6	E.Long	1.6
B.Short	4.4	D.Short	0.0		
B.Midlle	0.6	D.Midlle	2.8		
B.Long	0.0	D.Long	2.6		

Table 9. The membership values for sequence $(B.Short, C.Middle)$.

Client ID	Membership value of $(B.Short, C.Middle)$
1	0.0
2	0.8
3	0.0
4	0.6
5	0.8
6	0.0

Table 10. The fuzzy counts of the candidate 2-sequences in C_2.

Sequences	Count	Sequences	Count
$(B.Short, B.Short)$	1.0	$(B.Short, C.Middle)$	2.2
$(B.Short, D.Middle)$	1.2	$(B.Short, D.Long)$	0.8
$(C.Middle, B.Short)$	1.4	$(C.Middle, C.Middle)$	0.4
$(C.Middle, D.Middle)$	0.0	$(C.Middle, D.Long)$	0.0
$(D.Middle, B.Short)$	1.0	$(D.Middle, C.Middle)$	2.0
$(D.Middle, D.Middle)$	0.2	$(D.Middle, D.Long)$	0.2
$(D.Long, B.Short)$	2.0	$(D.Long, C.Middle)$	2.0
$(D.Long, D.Middle)$	0.6	$(D.Long, D.Long)$	0.4

STEP 14: Since L_2 is not null, $r = r + 1 = 2$. STEPs 12 to 14 are then repeated to find L_3. C_3 is first generated from L_2, and the sequence ($D.Long$, $B.Short$, $C.Middle$) is generated. Since its count is 0.4, smaller than 2.0, it is thus not put in L_3. L_3 is an empty set. STEP 15 then begins.

STEP 15: The maximally large sequences are output to web-site managers. In this example, the following four sequences are output as meta-knowledge concerning the given log data:

1. $B.Short \rightarrow C.Middle$,
2. $D.Middle \rightarrow C.Middle$,
3. $D.Long \rightarrow B.Short$,
4. $D.Long \rightarrow C.Middle$.

Note that if the transactions are run by the web mining algorithm proposed in [19], only the first two sequential patterns are found. The proposed algorithm here can thus generate a more complete set of sequential patterns than that in [19].

7 Experimental Results

The section reports on experiments made to show the time-completeness trade-off on fuzzy web-browsing mining. The experiments were implemented in C on a Pentium-III 700 personal computer. There were 90 web pages used in the experiments. 10000 client sequences were generated to comprise the data set. The number of browsed web pages in each client sequence was first randomly generated. The browsed web pages and their durations in each client sequence were then randomly generated.

The data set was run both by the proposed approach here (denoted Approach 1) and by the approach in [19] (denoted Approach 2). The relationships between numbers of browsing patterns mined and minimum support values for an average of 10 browsed pages in each client sequence are shown in Figure 2.

From Figure 2, it is easily seen that the numbers of browsing patterns by Approach 1 are greater than those by Approach 2. This is consistent with our discussion. The execution-time relationships for these two approaches are shown in Figure 3.

From Figure 3, it is easily seen that the execution times needed by Approach 1 are more than those needed by Approach 2. A trade-off thus exists between the time complexity and pattern completeness on fuzzy web-browsing mining.

8 Conclusions and Future Works

In this paper, we have proposed a fuzzy web-mining algorithm for processing web-server logs to discover fuzzy browsing patterns among them. The pro-

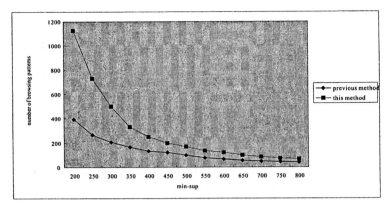

Fig. 2. The relationships between numbers of browsing patterns and minimum support values.

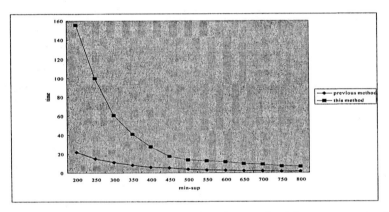

Fig. 3. The relationships between execution times and minimum support values.

posed algorithm can derive a more complete set of patterns than the method proposed in [19] although it needs more computation time. Trade-off thus exists between the computation time and the completeness of rules. Choosing an appropriate mining method thus depends on the requirements of the application domains.

Although the proposed method works well in data mining for quantitative values, it is just a beginning. There is still much work to be done in this field. Our method assumes that the membership functions are known in advance. In [15, 17], we proposed several fuzzy learning methods to automatically derive the membership functions. In the future, we will attempt to dynamically adjust the membership functions in the proposed web mining algorithm to avoid inappropriate choice of membership functions.

References

1. Agrawal R, Srikant R (1995) Mining sequential patterns. The Eleventh International Conference on Data Engineering 3–14
2. Blishun AF (1987) Fuzzy learning models in expert systems. Fuzzy Sets and Systems 22:57–70
3. Campos LM de and Moral S (1993) Learning rules for a fuzzy inference model. Fuzzy Sets and Systems 59:247–257
4. Chang RLP, Pavliddis T (1977) Fuzzy decision tree algorithms. IEEE Transactions on Systems. Man and Cybernetics 7:28–35
5. Chen MS, Park JS, Yu PS (1998) Efficient data mining for path taversal patterns. IEEE Transactions on Knowledge and Data Engineering 10:209–221
6. Chen L, Sycara K (1998) WebMate: A personal agent for browsing and searching. The Second International Conference on Autonomous Agents. ACM
7. Clair C, Liu C, Pissinou N (1998) Attribute weighting: a method of applying domain knowledge in the decision tree process. The Seventh International Conference on Information and Knowledge Management. 259–266
8. Clark P, Niblett T (1989) The CN2 induction algorithm. Machine Learning 3:261–283
9. Cohen E, Krishnamurthy B, Rexford J (1999) Efficient algorithms for predicting requests to web servers. The Eighteenth IEEE Annual Joint Conference on Computer and Communications Societies 1:284 –293
10. Cooley R, Mobasher B, Srivastava J (1997) Grouping web page references into transactions for mining world wide web browsing patterns. Knowledge and Data Engineering Exchange Workshop 2–9
11. Cooley R, Mobasher B, Srivastava J (1997) Web mining: information and pattern discovery on the world wide web. The Ninth IEEE International Conference on Tools with Artificial Intelligence 558–567
12. Delgado M, Gonzalez A (1993) An inductive learning procedure to identify fuzzy systems. Fuzzy Sets and Systems 55:121–132
13. Gonzalez A (1995) A learning methodology in uncertain and imprecise environments. International Journal of Intelligent Systems 10: 357–371
14. Graham I and Jones PL (1988) Expert Systems - Knowledge, Uncertainty and Decision. Chapman and Computing, Boston 117–158
15. Hong TP, Chen JB (1999) Finding relevant attributes and membership functions. Fuzzy Sets and Systems 103(3):389–404
16. Hong TP, Chen JB (2000) Processing individual fuzzy attributes for fuzzy rule induction. Fuzzy Sets and Systems 112(1):127–140
17. Hong TP, Lee CY (1996) Induction of fuzzy rules and membership functions from training examples. Fuzzy Sets and Systems 84:33–47
18. Hong TP, Kuo CS, Chi SC (1999) A data mining algorithm for transaction data with quantitative values. Intelligent Data Analysis 3(5): 363–376
19. Hong TP, Lin KY, Wang SL (2002) Mining linguistic browsing patterns in the world wide web. Soft Computing 6(5):329–336
20. Hong TP, Tseng SS (1997) A generalized version space learning algorithm for noisy and uncertain data. IEEE Transactions on Knowledge and Data Engineering 9(2):336–340
21. Hou RH, Hong TP, Tseng SS, Kuo SY (1997) A new probabilistic induction method. Journal of Automatic Reasoning 18:5-24

22. Kandel A (1992) Fuzzy Expert Systems. CRC Press, Boca Raton 8–19
23. Mamdani EH (1974) Applications of fuzzy algorithms for control of simple dynamic plants. IEEE Proceedings 1585–1588
24. Quinlan JR (1987) Decision tree as probabilistic classifier. The Fourth International Machine Learning Workshop. Morgan Kaufmann, San Mateo CA 31–37
25. Quinlan JR (1993) C4.5: Programs for Machine Learning. Morgan Kaufmann, San Mateo CA
26. Rives J (1990) FID3: fuzzy induction decision tree. The First International Symposium on Uncertainty, Modeling and Analysis 457–462
27. Wang CH, Hong TP, Tseng SS (1996) Inductive learning from fuzzy examples. The Fifth IEEE International Conference on Fuzzy Systems, New Orleans 13–18
28. Wang CH, Liu JF, Hong TP, Tseng SS (1999) A fuzzy inductive learning strategy for modular rules. Fuzzy Sets and Systems 103(1):91–105
29. Weber R (1992) Fuzzy-ID3: a class of methods for automatic knowledge acquisition. The Second International Conference on Fuzzy Logic and Neural Networks, Iizuka Japan 265–268
30. Yuan Y, Shaw MJ (1995) Induction of fuzzy decision trees. Fuzzy Sets and Systems 69:125–139
31. Zadeh LA (1988) Fuzzy logic. IEEE Computer 83–93
32. Zimmermann HJ (1991) Fuzzy Set Theory and Its Applications. Kluwer Academic Publisher, Boston

A Fuzzy Logic Approach for Content-Based Audio Classification and Boolean Retrieval

Mingchun Liu, Chunru Wan, and Lipo Wang

School of Electrical and Electronic Engineering
Nanyang Technological University
Block S2, 50 Nanyang Avenue, Singapore 639798
p147508078,ecrwan,elpwang@ntu.edu.sg

Summary. Since the invention of fuzzy sets and maturing of the fuzzy logic theory, fuzzy logic systems have been widely applied to various fields, such as fuzzy controller, data mining, and so on. New potential areas using fuzzy logic are also being explored with the emergence of other technologies. One booming technology today is the Internet, due to its fast growing number of users and rich contents. With huge data storage and speedy networks becoming available, multimedia contents like image, video, and audio are fast increasing. In order to search and index these media effectively, various content-based multimedia retrieval systems have been studied.

In this chapter, we introduce a fuzzy logic approach for hierarchical content-based audio classification and boolean retrieval, which is intuitive due to the fuzzy nature of human perception of audio, especially audio clips of mixed types. The fuzzy nature of audio search lies in the facts that (1) both the query and target are approximations of the user's memory and desire and (2) exact matching is sometimes impossible or impractical. Therefore, fuzzy logic systems are a natural choice in audio classification and retrieval.

The fuzzy tree classifier is the core of the hierarchical content-based audio classification. At the beginning, audio features are extracted for audio samples in the database. Proper features are then selected and used as input to a constructed fuzzy inference system (FIS). The outputs of the FIS are two types of hierarchical audio classes. The membership functions and rules are derived from the distributions of the audio features. Non-speech and music sounds are discriminated by the FIS in the first hierarchy. Secondly, music and speech are separated. One particular sound, the telephone ring, has also been recognized in this level. In the prototype system, the classification ability of up to fourth level has been explored. Hence we can use multiple FISs to form the 'fuzzy tree' for retrieval of different types of audio clips. With this approach, we can classify and retrieve generic audios using fewer features and less computation time, compared to other existing approaches.

As for retrieval, the existing content-based audio retrieval systems usually adopt the query-by-example mechanism to search for desired audio files. However, only one single audio sample often cannot express the user's needs adequately. To overcome this problem, more audio files can be chosen as queries provided by the user or through feedback during searching. Correspondingly, we present a different scheme to handle content-based audio retrieval with multi-queries. The multiple queries are linked by boolean operators and thus it can be treated as a boolean search problem. We build a framework to solve the three basic boolean operators known as AND, OR, and NOT, with concepts adopted from fuzzy logic. Experiments have shown that boolean search can be helpful in audio retrieval.

1 Introduction

Traditional search engines such as Google and Yahoo can provide a portal for web surfers to find their interested web pages. Yet, commercial search engines for multimedia databases, especially for audios, are lacking. Some search engines do provide some search ability on multimedia contents, but most of these systems are based on the surrounding texts or titles of the multimedia data. However, users can benefit from the ability to directly search these media, which contain rich information but could not be precisely described by text. Hence, content-based indexing and retrieval technologies are the first crucial step towards building such multimedia search engines. In recent years, research has been conducted on content-based audio classification and retrieval, as well as in other relevant fields, such as audio segmentation, indexing, browsing and annotation.

Generally, audio can be categorized into three major classes: speech, music, and sound. Different techniques have been employed to process these three types of audios individually. Speech signals are the best studied. With automatic speech recognition systems becoming mature, speech and spoken document retrievals are often carried out by transforming the speeches into texts. Traditional text retrieval strategies are then used [1], [2], [3]. Music retrieval is sometimes treated as a string matching problem. In [4], a new approximate string matching algorithm is proposed to match feature strings, such as melody strings, rhythm strings, and chord strings, of music objects in a music database. Besides speech and music, general sounds are the third major type of audios. Some research has been devoted to classification of this kind of audios, and others focus on even more specific domains, such as classification of piano sounds [5] and ringing sounds [6].

In spite of different techniques applied in the audio classification and retrieval process, the underlying procedure is similar, which can be divided into three major steps: audio feature extraction, classifier mapping, and distance ranking.

The first step towards these content-based audio database systems is to extract features from sound signals. Based on the features extracted, various classifiers can then be used for sound classification. In [7], a multidimensional Gaussian maximum a posteriori (MAP) estimator, a Gaussian mixture model (GMM) classifier, a spatial partitioning scheme based on a k-d tree, and a nearest neighbor classifier were examined in depth to discriminate speech and music. In [8], a threshold-based heuristic rule procedure was developed to classify generic audio signals, which was model-free. The Hidden Markov model (HMM) was used in [9] to classify TV programs into commercial, basketball, football, news and weather, based on audio information.

Once an audio has its label, it can be indexed and annotated for browsing and retrieval. Contrary to using keywords in queries for text retrieval, examples are used in queries for audio files. Usually, the similarities between the audio samples in the database and the query example are calculated, a distance-ranking list is given as the retrieval result.

Content-based audio retrieval can be a useful feature towards a multimedia search engine. Wold et al built a general audio classification and retrieval system which led the research along this direction [10]. In that system, sounds are reduced to perceptual and acoustical features, which let users search or retrieve sounds by different kinds of query. A new pattern classification method called the nearest feature line (NFL) was presented for the same task and experiments were carried out based on the same database [15]. The resulting system achieved lower error rate. With increasing audio types being explored, an online audio classification and segmentation system is presented [11]. Outlines of further classification of audio into finer types and a query-by-example audio retrieval system on top of the coarse classification are also introduced.

There also exists some research in audio classification and retrieval using fuzzy logic. In [12], a new method for multilevel speech classification based on fuzzy logic has been proposed. Through simple fuzzy rules, their fuzzy voicing detector system achieves a sophisticated speech classification, returning a range of continuous values between extreme classes of voiced/unvoiced. In classification of audio events in broadcast news [13], when fuzzy membership functions associated with the features are introduced, the overall accuracy of hard threshold classifier can be improved by 4.5% and achieved 94.9%. All these related work has demonstrated the ability of fuzzy logic to enhance classification performance and thus given more or less hints for us to conduct our research of audio classification and retrieval with fuzzy inference systems.

In this chapter, we adopt a fuzzy logic approach and build a fuzzy-tree hierarchical classifier named 'fuzzy-tree', for content-based audio classification. During the searching process, we further propose a fuzzy expert system to deal with AND, OR, and NOT boolean queries.

The rest of the chapter is organized as follows. Audio feature extraction and normalization, which are prerequisite steps in a content-based system, are discussed in Section 2. The proposed fuzzy inference system for audio classification is described in Section 3. The experimental results of the fuzzy-tree classifier and its application in audio retrieval are presented in Section 4. Then, the boolean search algorithm is proposed and different fuzzy concepts and rules are illustrated in Section 5. In Section 6, various experiments have been carried out to test the performance of the proposed boolean search. Finally, conclusions are given in Section 7.

2 Audio Feature Extraction and Normalization

In order to classify audios automatically, features are to be extracted from raw audio data source at the beginning. The extracted feature vectors are then normalized for classification and indexing. The audio database being classified in this chapter is described in Table 1. It is a common audio database as in [10], [15] and [14]. The lengths of these files range from about half a second to less than ten seconds. They are sorted into three major categories: speech, music and sound. The database has 16 classes: two from speech (female and male speech), seven from music (percussion, oboe, trombone, cello, tubular-bell, violin-bowed, violin-pizzicato), and seven from other sounds including telephone ring. Fuzzy logic will be applied to hierarchically classify the audio into their corresponding classes. The inputs to the fuzzy inference system (FIS) are some selected features.

Table 1. Structure of database I

Class name	Number of files	Class name	Number of files
1.Speech	53	Violin-pizzicato(9)	40
Female(1)	36	3.Sound	62
Male(2)	17	Animal(10)	9
2.Music	299	Bell(11)	7
Trombone(3)	13	Crowds(12)	4
Cello(4)	47	Laughter(13)	7
Oboe(5)	32	Machines(14)	11
Percussion(6)	102	Telephone(15)	17
Tubular-bell(7)	20	Water(16)	7
Violin-bowed(8)	45	Total	414

We extract features from the time, frequency, and coefficient domain. They are obtained by calculating the mean and standard deviation of frame-level characteristics. These characteristics are computed from 256 samples per frame, with 50 % overlap between two adjacent frames from hamming-windowed original sound.

Time domain features include RMS (root mean square), ZCR (zero-crossing ratio), VDR (volume dynamic ratio) and silence ratio. Frequency domain features include frequency centroid, bandwidth, four sub-band energy ratios, pitch, salience of pitch, spectrogram, first two formant frequencies, and formant amplitudes. The coefficient features are the first 13 orders of MFCCs (Mel-Frequency Cepstral Coefficients) and LPCs (Linear Prediction Coefficients). A summary of the features are list in Table 2.

Table 2. Structure of 84 extracted features

1 Time domain (6 features)	Mean and standard deviation of volume root mean square (RMS), zero-crossing ratio; volume dynamic ratio (VDR) and silence ratio.
2 Frequency domain (26 features)	Mean and standard deviation of frequency centroid, bandwidth, four sub-band energy ratios, pitch, salience of pitch, spectrogram, first two formant frequencies and amplitudes.
3 Coefficient domain (52 features)	Mean and standard deviation of first 13 orders of MFCCs (Mel-Frequency Cepstral Coefficients) and LPCs(Linear Prediction Coefficients).

2.1 Time Domain Features

Time domain features include RMS (root mean square), ZCR (zero-crossing ratio), VDR (volume dynamic ratio) and silence ratio:

- **RMS:** The measure of loudness of the frame. This feature is unique to segmentation since changes in loudness are important cues for new sound events.

$$RMS_j = \sqrt{\frac{1}{N} \sum_{m=1}^{N} x_j^2(m)} \tag{1}$$

where $x_j(m)(m = 1, 2, \cdots, N)$ is jth frame of windowed audio signal of length N. N is the number of samples in each frame. We have set N to be 256 in all of our experiments below.

- **Zero-Crossing Ratio:** A zero-crossing is said to occur if successive samples have different signs. The zero-crossing ratio is the number of the time-domain zero-crossings and total number of samples in a frame.

$$\mathbf{Z}_j = \frac{1}{2N} \sum_m |sgn[x_j(m)] - sgn[x_j(m-1)]| \tag{2}$$

where $sgn[x(n)] = \begin{cases} 1 & , \quad x(n) \geq 0 \\ -1 & , \quad x(n) < 0 \end{cases}$.

- **VDR:** It is the difference of maximum and minimum RMS normalized by the maximum RMS of the frame audio signal. The magnitude of VDR is dependent on the type of the sound source.

$$VDR = \frac{Max_j(RMS_j) - Min_j(RMS_j)}{Max_j(RMS_j)} \tag{3}$$

- **Silence Ratio:** It is the ratio of silent frames (determined by a preset threshold) and the entire frames. Here, a frame is said to be silent if the frame RMS is less than 10% of mean of RMS of the files.

$$SR = \frac{Number\ of\ Silence\ Frame}{Total\ Number\ of\ Frames} \qquad (4)$$

2.2 Frequency Domain Features

The features used in frequency domain include frequency centroid, bandwidth, four sub-band energy ratios, pitch, salience of pitch, spectrogram, first two formant frequencies, and formant amplitudes.

- **Frequency Centroid (Brightness):** It represents the balancing point of the spectrum.

$$\omega_c j = \frac{\int_0^{\omega_0} \omega |X_j(\omega)|^2 d\omega}{\int_0^{\omega_0} |X_j(\omega)|^2 d\omega} \qquad (5)$$

where $|X_j(\omega)|^2$ is the power spectrum of $x_j(m)$ and ω_0 is the half sampling frequency.
- **Bandwidth:** It is the magnitude-weighted average of the difference between the spectral components and the frequency centroid.

$$B_j = \sqrt{\frac{\int_0^{\omega_0} (\omega - \omega_c)|X_j(\omega)|^2 d\omega}{\int_0^{\omega_0} |X_j(\omega)|^2 d\omega}} \qquad (6)$$

- **Sub-Band Energy Ratio:** The frequency spectrum is divided into 4 sub-bands with intervals $[0, \frac{\omega_0}{8}], [\frac{\omega_0}{8}, \frac{\omega_0}{4}], [\frac{\omega_0}{4}, \frac{\omega_0}{2}], [\frac{\omega_0}{2}, \omega_0]$. The sub-band energy ratio is measured by $\frac{P_k}{P}$, where $P_k = \int_{L_k}^{H_k} |X_j(\omega)|^2 d\omega$, $P = \int_0^{\omega_0} |X_j(\omega)|^2 d\omega$, L_j and H_j are lower and upper bound of sub-band k. Sub-band energy ratios, when used together, reveal the distribution of spectral energy over the entire frame.
- **Pitch:** Pitch refers to the fundamental period of a human speech waveform. We compute the pitch by finding the time lag with the largest autocorrelation energy. It is an important parameter in the analysis and synthesis of speech.
- **Salience of Pitch:** It is the ratio of the first peak (pitch) value and the zerolag value of the autocorrelation function, defined as $\frac{\phi_j(P)}{\phi(0)}$.

$$\phi_j(P) = \sum_{m=-\infty}^{\infty} x_j(m)x_j(m - P), \phi(0) = \sum_{m=-\infty}^{\infty} x^2(m)^2 \qquad (7)$$

where $\phi_j(P)$ is the pitch value of the autocorrelation function, $\phi(0)$ is the zerolag value of the autocorrelation function.

- **Spectrogram:**Spectrogram splits the signal into overlapping segments, windows each segment with the hamming window and forms the output with their zero-padded, N points discrete Fourier transforms. Thus the output contains an estimate of the short-term, time-localized frequency content of the input signal. We compute the statistics of the absolute value of the elements of spectrogram matrix as features.
- **First two Formants and amplitudes:** Formant is caused by resonant cavities in the vocal tract of a speaker. The first and second formants are most important.

2.3 Coefficient Domain Features

The MFCC (mel-frequency cepstral coefficients) and LPC (linear prediction coefficients) coefficients, which are widely used in speech recognition, are also adopted for classification of general sounds.

- **Mel-Frequency Cepstral Coefficients:** These are computed from the FFT power coefficients. We adopt the first 13 orders of coefficients.
- **Linear Prediction Coefficients:** The LPC coefficients are a short-time measure of the speech signal, which describe the signal as the output of an all-pole filter. The first 13 orders of LPC parameters are calculated.

2.4 Feature Normalization

Normalization can ensure that contributions of all audio feature elements are adequately represented. Each audio feature is normalized over all files in the database by subtracting its mean and dividing by its standard deviation. The magnitudes of the normalized features are more uniform, which keeps one feature from dominating the whole feature vector. Then, each audio file is fully represented by its normalized feature vector.

3 Fuzzy Inference System

There are several important issues in building a Fuzzy Inference System (FIS), such as selecting the right features as inputs of the system, constructing proper membership functions and rules, and tuning parameters to achieve a better performance.

3.1 Selecting Features as Inputs

In order to select appropriate features as inputs to the FIS from the extracted ones [16], we use a simple nearest neighbor (NN) classifier and a sequential forward selection (SFS) method to choose the appropriate features. The entire data set is divided into two equal parts for training and testing the NN classifier. Firstly, the best single feature is selected based on classification accuracy it can provide. Next, a new feature, in combination with the already selected feature, is added in from the rest of features to minimize the classification error rate, in order to find the combination of two features that leads to the highest classification accuracy. Our objective is to use as few features as possible to achieve a reasonable performance. Experiments show that using one or two features is adequate to do the hierarchical classification at each level. These features are thus chosen as inputs to the FIS. Through experiments, a hierarchical 'fuzzy-tree' is constructed by combining up to 4 levels FIS together, shown in Figure 1. The proper features for each FIS are listed below in Table 3. Note that, *std* in Table 3 refers to 'standard deviation'.

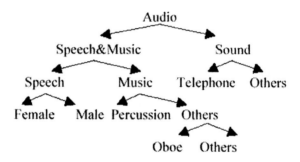

Fig. 1. The hierarchical fuzzy tree for retrieval for retrieval

Table 3. Fuzzy classifiers and their input features

FIS Classifiers	Features
sound and others	mean Spectrogram and mean of third MFCC coefficient
speech and music	mean spectrogram and std of pitch salience ratio
telephone ring and others	mean salience
female and male speech	mean pitch and mean pitch salience
percussion and others	mean pitch salience and mean of first MFCC coefficient
oboe and others	mean zero crossing ratio and std of centroid

For example, mean spectrogram and mean of the third MFCC coefficient are inputs for sound recognition and mean spectrogram and standard deviation of pitch salience ratio are used for discriminating speech and music. From

Table 3, we observed that the features of spectrogram and pitch salience are very useful because they have been used as input in various FIS.

3.2 Membership Function and Rule Construction

In order to demonstrate the design procedure of fuzzy classifiers, we take the following two examples: One is to classify male and female speech in the third hierarchy, the other is to identify one particular sound, the telephone ring, from a sound collection in the second hierarchy. In the female and male speech classifier, the two feature (pitch and pitch salience) histograms of the two classes are shown in Figure 2(a) and Figure 3(a), respectively. Each histogram is normalized by its peak value. After determining the inputs, the key to constructing the fuzzy classifier is to design the membership function and extract rules. In fact, the membership functions of each input and output, as well as the rules, can be derived from simulating the feature distributions. We chose Gaussian membership functions, which is fully parameterized by the mean and the standard deviation. We calculate these parameters directly from the statistics of the features among the whole data source. We use 'small' and 'large' to denote their membership according to their class distribution. The resulting simplified Gaussian membership functions simulating the feature distributions are shown in Figure 2(b) and Figure 3(b). Another two Gaussian membership functions are chosen for output, shown in Figure 4. One mean is zero and another mean is one. Both have same standard deviation that it makes equal probabilities at center of their distributions. An overview of the fuzzy classifier for discriminating female and male speech are given in Figure 5. The rules in the FIS are listed below.

- If (*Mean Pitch*) is *small* AND (*Mean Pitch Salience ratio*) is *small* Then (*Type is male speech*)

- If (*Mean Pitch*) is *large* AND (*Mean Pitch Salience ratio*) is *large* Then (*Type is female speech*)

Another example is to identify one special sound, the telephone ring among a sound collection. Because the telephone ring sounds differently from other sounds, it can be correctly classified with 100% accuracy. The input feature histogram is shown in Figure 6(a). The simplified Gaussian membership functions simulating the feature distributions is shown in Figure 6(b). The Gaussian membership functions for output is shown in Figure 7. The whole FIS for identification of telephone ring is given in Figure 8. The rules for the FIS are as follows.

- If (*Mean Pitch salience ratio*) is *large* Then (*Type is telephone ring*)

- If (*Mean Pitch salience ratio*) is *small* Then (*Type is others*)

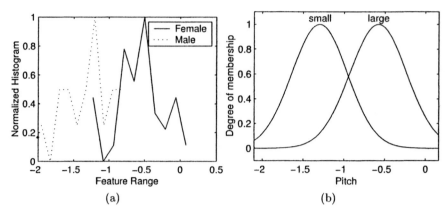

Fig. 2. (a)The feature distribution of mean pitch for female and male, and (b) The Gaussian membership function simulating the feature distribution of mean pitch for female and male.

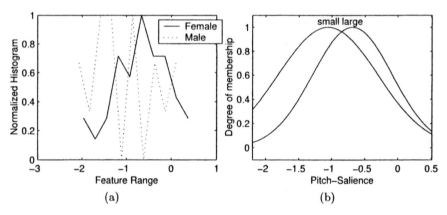

Fig. 3. (a)The feature distribution of mean pitch salience ratio for female and male, and (b) The Gaussian membership function simulating the feature distribution of mean pitch salience ratio for female and male.

Similarly, the rules for the FIS to distinguish music and speech from sound are:

- If (*Mean Spectrogram*) is *small* AND (*mean of the third MFCC coefficient*) is *large* Then (*Type is music and speech*)

- If (*Mean Spectrogram*) is *large* AND (*mean of the third MFCC coefficient*) is *small* Then (*Type is sound*)

The rules for the FIS to classify music and speech are:

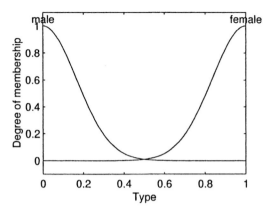

Fig. 4. The Gaussian membership function simulating the output for female and male

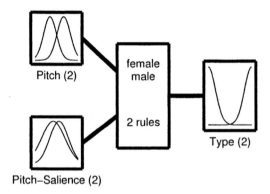

System female&male: 2 inputs, 1 outputs, 2 rules

Fig. 5. The FIS input-output diagram for female and male classifier

- If (*Mean Spectrogram*) is *small* or (*Standard deviation of pitch salience ratio*) is *small* Then (*Type is music*)

- If (*Mean Spectrogram*) is *large* or (*Standard deviation of pitch salience ratio*) is *large* Then (*Type is speech*)

The rules for the FIS to identify percussion from non-percussion musical instrumental sounds are:

145

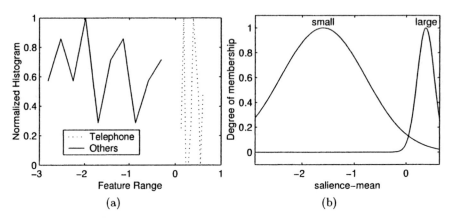

Fig. 6. (a)The feature distribution of mean pitch salience ratio for telephone ring and other sounds, and (b) The Gaussian membership function simulating the feature distribution of pitch salience ratio for telephone ring and other sounds.

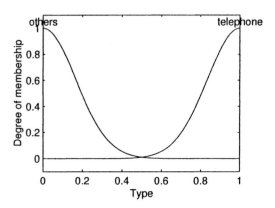

Fig. 7. The Gaussian membership function simulating the output for telephone ring and other sounds

- If (*Mean Pitch salience ratio*) is *small* AND (*mean of first MFCC coefficie* is *small* Then (*Type is* percussion)

- If (*Mean Pitch salience ratio*) is *large* AND (*mean of first MFCC coefficie* is *large* Then (*Type is* non-percussion)

The rules for the FIS to differentiate oboe from music instruments are:

- If (*Mean zero crossing ratio*) is *large* AND (*standard deviation of centroid*) is *small* Then (*Type is oboe*)

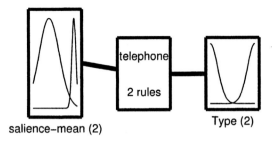

salience−mean (2)

telephone
2 rules

Type (2)

System telephone: 1 inputs, 1 outputs, 2 rules

Fig. 8. The FIS input-output diagram for telephone ring classifier

- If (*Mean zero crossing ratio*) is *small* AND (*standard deviation of centroid*) is *large* Then (*Type is others*)

Note that, we first use both AND and OR connector connecting inputs if the inputs are more than 1. Then, we decide the relation with the higher classification accuracy. During the experiments, we found that introducing more inputs doesn't improve performance greatly. In some cases, the performance can even decline. For example, the accuracy of oboe classifier with 1, 2, 3 inputs are 84%, 94% and 93% respectively. Therefore, we design our classifier with one or two inputs.

3.3 Tuning the FIS

Although the fuzzy inference systems are thus constructed completely, there are ways to improving the performance, for example, by tuning parameters of those membership functions, choosing other types of membership function corresponding to the feature distribution, or using neural networks to train the membership functions for a closer approximation. Since those features selected by the sequential forward selection method are sub-optimum inputs, we may also try other combinations of features as input to improve accuracy.

4 FIS Classification and Its Application in Audio Retrieval

4.1 FIS Classification Results

In the integrated 'fuzzy-tree' classification system, shown previously in Figure 1, all classifications can be done hierarchically. The target level depend on user's interest. For example, when an audio is submitted to the system for classification, the first level FIS can distinguish speech and music from sound. Then, if the result is music and speech, the second level FIS can further tell whether it is music or speech. If the result is sound, the second level FIS can detect whether it belongs to one particular sound, such as telephone ring. Other FIS classifiers can recognize female and male speech, identify percussion and oboe music instruments, for examples. With more domain knowledge collected, we may discover new features and new rules which fit for identifying sounds such as thunder, laughter, applause and so on, or further probe their semantic meanings. In addition, other music types can be recognized from instrument family by studying their vibration characteristics. Experiments have been done hierarchically to obtain the performance of all these fuzzy classifiers. At the first level of the fuzzy tree, each audio file is used as input to the fuzzy classifier. At the second level of the fuzzy tree, the experiments are conducted based on the subset of the audio files. For example, 352 speeches and musics are submitted to music and speech classifier, 62 sounds are submitted to telephone ring detector. Further, 53 speech files, 299 and 197 music files are tested for female and male classifier, percussion detector and oboe detector respectively. The Percussion is firstly distinguished from the rest of music instrument because of its inharmonic nature. All classification results are summarized in Table 4.

Table 4. Classification performance

FIS classifier	Classification Accuracies
Sound and Others	80%
Music-Speech	92%
Telephone ring	100%
Female-male	89%
Percussion	81%
Oboe	94%

4.2 Fuzzy-tree for Audio Retrieval

Content-based audio search and retrieval can be conducted as follows. When a user inputs a query audio file and requests to find relevant files, both the query and each audio file in the database are represented as feature vectors. A measure of the similarity between the query feature vector and a stored feature vector is evaluated and a list of files based on the similarity are fed back to the user for listening and browsing. The user may refine the query to get audios more relevant to his or her interest by feedbacks. The performance of retrieval is measured by precision and recall defined as follows:

$$Precision = \frac{Relevant\ Retrieved}{Total\ Retrieved} \tag{8}$$

$$Recall = \frac{Relevant\ Retrieved}{Total\ Relevant} \tag{9}$$

Sometimes, the average precision is used as another measurement of retrieval performance, which refers to an average of precision at various points of recall. Precision indicates the quality of the answer set, while recall indicates the completeness of the answer set. In an ideal situation, precision is always 1 at any recall point. The fuzzy-tree architecture as shown previously can be helpful for retrieval. When a query input is presented, the direct search may result in mixture types of audio clips retrieved. If we firstly classify the query into one particular node of the fuzzy tree, we can then search relevant files only in that subspace instead of the whole database. For example, various audio files can appear in the search results of a speech query. If we can firstly classify it to one subset like speech and music category, lots of irrelevant sounds can be discarded before search begins. Thus, the precision will increase and the searching time will decrease. If the classification is wrong, we can search in other branches of the fuzzy tree with user's feedback. Then, a Euclidean distance method is adopted to select the most similar samples in the database within the selected class. When the database grows, new classes can be added to the tree. Only links between the new class and its immediate upper level are updated, with the rest of the tree unchanged.

5 Boolean Search Using Fuzzy Logic

In the existing content-based audio retrieval systems, a single query example is usually considered as input to the audio search engine. However, this single audio sample often cannot express the user's needs sufficiently and adequately. In many cases, even the user cannot provide more examples at hand. However, additional queries can be generated through feedback in the searching process. The multiple query examples can be linked by boolean operators and it thus can be treated as a boolean search problem. While in the traditional textual documents retrieval, such boolean query is commonly used. With these observed, we propose a scheme to handle boolean query in the audio retrieval domain. We build a framework to solve the three basic boolean operators known as AND, OR, and NOT, with concept adopted from fuzzy logic. Because of the similarities between boolean query and fuzzy logic, we proposed a fuzzy expert system which can translate the boolean query for retrieval purpose.

5.1 Multi-example Query

When a user wants to retrieve desired audio documents, the system usually requires a query example as input to begin the search process. A similarity measurement such as Euclidean distance between the query and sample audio files is computed. Then, lists of files based on the similarity are displayed to the user for listening and browsing. As mentioned earlier, it is usually incomplete to use one query to express the user's needs. Therefore, the boolean query can represent user's request more adequately by combining multiple query examples.

5.2 The Fuzzy Logic and Boolean Query

A boolean query has a syntax composed of query examples and boolean operators. The most commonly used operators, given two basic queries q1 and q2, are as follows.

- AND, where the query (q1 AND q2) selects all documents which satisfy both q1 and q2.
- OR, where the query (q1 OR q2) selects all documents which satisfy either q1 or q2.
- NOT, the query (q1 AND (NOT q2)) selects all documents which satisfy q1 but not q2. In case (NOT q2), all documents not satisfying q2 should be delivered, which may retrieve a huge amount of files and is probably not what the user wants.

These three boolean operations corresponding to intersection, union and complement operations in fuzzy logic system. Let U be a universal set, A and B be two fuzzy subsets of U, and $\bar{A}$ be the complement of A relative to U, and u be an element of U. These operations can be defined as:

- Intersection: $\mu_{(A) \bigcap (B)} = min(\mu_A(u), \mu_B(u))$
- Union: $\mu_{(A) \bigcup (B)} = max(\mu_A(u), \mu_B(u))$
- Complement: $\mu_{\bar{A}}(u) = 1 - \mu_A$

where $\mu(\cdot)$ is the membership function.

5.3 Similarity Measurement and Membership Function

In order to utilize the fuzzy expert system for audio retrieval, we first calculate the Euclidean distance between the query and samples in the database and then define the similarity measurement as follows.

$$Dist(q, d_i) = \sqrt{\sum_{j=1}^{N} (q_j - d_{ij})^2} \qquad (10)$$

$$Sim(q, d_i) = \frac{1}{Dist(q, d_i) + 1} \qquad (11)$$

where q is the query audio feature vector, d_i is the feature vectors of the ith file in the database. The q_j is the jth element of the feature vector q, d_{ij} is the jth element of the feature vector d_i. Since the distance $Dist(q, d_i)$ ranges from $[0, \infty)$, thus, the similarity $Sim(q, d_i)$ ranges from $(0, 1]$. Then, we can use the similarity as membership function for the selected file in the database.

5.4 The Fuzzy Expert System for Boolean Query

Suppose a general boolean query is proposed for searching audio files as following.

"Find documents which sound similar to ((q1 and q2) and (not q3)) or q4"

We decomposed the boolean query into a fuzzy expert system as follows.

- Rule 1: If $Sim(q1, d_i)$ AND $Sim(q2, d_i)$ AND (not $Sim(q3, d_i)$) Then file similarity is high
- Rule 2: If $Sim(q4, d_i)$ Then file similarity is high

The method of the decomposition is that we group the AND and NOT logic into one rule and OR logic into another. During the process, the AND boolean is always performed before OR boolean. Then, $Max[(rule1), (rule2)]$ is used to combined each output to form the final similarity for sorting.

5.5 The Retrieval Procedure

The general inference process of a fuzzy expert system proceeds in following steps.

1. Firstly, we calculate distance and similarity defined in Section 5.3 to determine the degree of truth for each rule premise.
2. Secondly, we conduct query decomposition and AND, OR rule inference introduced in Section 5.4. This results in one fuzzy subset to be assigned to each output variable for each rule.
3. Thirdly, we maximize both decomposed rules to form a single fuzzy subset for each output variable and convert the fuzzy output to a crisp number.

6 Boolean Query Experiments

We conduct boolean query experiments on two databases. The first one has been described earlier in FIS classification. The second database consists samples from seven classes: music(10), speech(32), sound(12), speech-music mixed(16), speech-sound mixed(18), sound-music mixed(14), and mixture(14) of all the three audio classes. The number in each bracket is the number of samples in each class. All the files are extracted clips ranging from several seconds from the movie 'Titanic' with a total number of 117. During the experiments, single query and boolean queries are submitted to search for required relevant audio documents. The performance are evaluated and compared between single queries and boolean queries.

6.1 Experiments on AND Boolean Queries from Same Class

The AND boolean operator is the most frequently used boolean search, which normally link two examples from same class to have a more precise representation of the user's query needs. The experiment is conducted on the first database. The result of one example of such an AND boolean query against the results of the two individual query is shown in Table 5 and Figure 9. We only list the first 15 ranked files in Table 5, because normally users only browse files listed at the top. In Table 5, q1 and q2 are both audio files from class Violinbowed. From Table 5, we can see that by using a single query, there are 8 and 9 files in the same class as the query example, which is from the Violinbowed. By using the AND query formed by two examples, there are 12 relevant retrieved. The average precision of the three retrieval are 0.32, 0.35 and 0.39 respectively. The recall and precision curve in Figure 9 also shows that the AND query performs generally better than the two individual queries.

6.2 Experiments on AND Boolean Queries from Different Classes

Sometimes, the AND boolean operator can also link two queries from different classes. The result of one example of such an AND boolean query against the results of the two individual queries is shown in Table 6. This experiment is conducted on the second database. Here, We only list the first 10 ranked files. In Table 6, q1 and q2 are audio files from class speech and sound respectively. It is shown that if the two examples linked by AND boolean are from different audio classes, the retrieved samples contain both characteristics of the two classes. This is due to the fact that for AND boolean, only the files both similar to these two query examples will appear at the top. In this way, some semantic searching could be explored.

Table 5. AND Boolean Query Results from Same Class

Rank	q1	q2	q1 and q2
1	Violinbowed	Violinbowed	Violinbowed
2	Violinbowed	Violinbowed	Violinbowed
3	Violinbowed	Violinbowed	Cellobowed
4	Cellobowed	Cellobowed	Violinbowed
5	Altrotrombone	Violinbowed	Violinbowed
6	Violinbowed	Violinbowed	Violinbowed
7	Violinbowed	Violinbowed	Altrotrombone
8	Cellobowed	Violinbowed	Violinbowed
9	Cellobowed	Oboe	Violinbowed
10	Oboe	Cellobowed	Violinbowed
11	Violinbowed	Cellobowed	Violinbowed
12	Violinbowed	Violinbowed	Violinbowed
13	Violinpizz	Violinbowed	Violinbowed
14	Altrotrombone	Cellobowed	Violinbowed
15	Violinbowed	Oboe	Oboe
Relevant	8	9	12

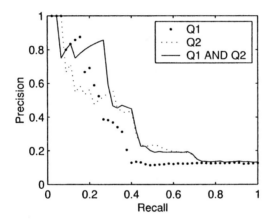

Fig. 9. Precision and Recall curve of AND boolean query.

6.3 Experiments on OR Boolean Queries

In the third experiment, we test the result of OR boolean query linking female and male speech based on the first database again. The result in Table 7 shows that samples from both query classes could appear in the OR boolean query. This is due to the fact that in OR query, the files similar to any of the two query examples may rank at the top.

Table 6. AND Boolean Query Results from different Classes

Rank	q1	q2	q1 and q2
1	Speech	Sound	Speech-sound
2	Speech	Sound	Speech-sound
3	Speech-sound	Sound	Speech-sound
4	Speech	Sound-music	Speech-sound
5	Speech	Sound	Sound-music
6	Speech	Sound-music	Speech-sound
7	Speech	Speech-music	Speech-sound
8	Speech	Sound-music	Speech-sound
9	Speech	Sound	Speech-sound
10	Speech	Music	Mixture

Table 7. OR Boolean Query Results

Rank	q1	q2	q1 or q2
1	Female	Male	Female
2	Female	Male	Male
3	Male	Female	Female
4	Male	Female	Male
5	Female	Female	Male
6	Female	Machines	Female
7	Male	Animal	Female
8	Female	Female	Female
9	Female	Female	Machines
10	Female	Female	Male

6.4 Experiments on Mixed Boolean Queries

A mixed boolean query which contains both AND and OR operators from different classes is shown in Table 8. This experiment is conducted on the second database again. The q1 , q2, and q3 are from speech, music, and sound, respectively. From Table 8, we can see that samples which satisfy AND query are found firstly, then it merges with the OR boolean query to obtain the final results. Note that, in the NOT query, such as q1 NOT q2, the undesired sample q2 can be moved to bottom in the retrieve list by the NOT boolean query each time, though the result is not shown here.

7 Conclusion

In this chapter, we propose a fuzzy inference system for audio classification and retrieval, as a first step towards a multimedia search engine for the Internet. The benefits of the fuzzy classifier lie in the facts that no further training

Table 8. Mixed Boolean Query Results

Rank	q1	q2	q3	q1 and q2	q1 and q2 or q3
1	Speech	Music	Sound	speech	Sound
2	Speech	Music	Sound	Speech-music	Sound
3	Speech	Sound-Music	Speech-music	Speech-music	Speech-music
4	Speech	Mixture	Sound-music	Speech-music	Sound-music
5	Speech	Music	Sound	Music	Speech
6	Speech	Sound	Sound	Sound	Sound
7	Speech	Sound	Sound-music	Speech-sound	Sound
8	Speech	Speech-music	Sound-music	Mixture	Speech-music
9	Speech	Music	Sound	Speech-sound	Speech-music
10	Speech	Sound	Sound	Speech-Sound	Speech-music

is needed once the fuzzy inference system is designed. Thus, classification can be performed very quickly. In addition, when the database grows, new classes can be added to the fuzzy tree. Only links between that class and its immediate upper level are required to be updated, with the rest of the tree unchanged. With this architecture, fast online web applications can be built. Future work along this direction is to use neural networks to train the parameters to obtain better membership functions, and to explore new features and rules to classify various audios with the so-called 'fuzzy tree' for hierarchical retrieval. In addition, we proposed a general method based on fuzzy expert system to handle boolean query in audio retrieval. The boolean query can be used in both direct boolean search and user's feedback. Some intelligence or semantics can be also discovered through the searching process thus the gap between the subjective concepts and objective features can be narrowed. In this way, we hope not only to enhance the retrieval performance but also to enhance searching ability. The boolean search algorithm can be used in image and video retrieval, as well as the user feedback scenario.

References

1. Makhoul J, Kubala F et al. (2000) Speech and language technologies for audio indexing and retrieval code. In: Proceedings of the IEEE, Volume: 88 Issue: 8, Aug 2000, pp: 1338 -1353
2. Viswanathan M, Beigi H.S.M et al. (1999) Retrieval from spoken documents using content and speaker information. In: ICDAR'99 pp: 567 -572
3. Gauvain J.-L, Lamel L (2000) Large-vocabulary continuous speech recognition: advances and applications. In: Proceedings of the IEEE, Volume: 88 Issue: 8, Aug 2000, pp: 1181 -1200
4. Chih-Chin Liu, Jia-Lien Hsu, Chen A.L.P (1999) An approximate string matching algorithm for content-based music data retrieval. In: IEEE International Conference on Multimedia Computing and Systems, Volume: 1, 1999, pp: 451 -456

5. Delfs C, Jondral F (1997) Classification of piano sounds using time-frequency signal analysis. In: ICASSP-97, Volume: 3 pp: 2093-2096
6. Paradie M.J, Nawab S.H (1990) The classification of ringing sounds. In: ICASSP-90, pp: 2435 -2438
7. Scheirer E, Slaney M (1997) Construction and evaluation of a robust multifeature speech/music discriminator. In: ICASSP-97, Volume: 2, pp: 1331 -1334
8. Tong Zhang, C.-C. Jay Kuo (1999) Heuristic approach for generic audio data segmentation and annotation. In: ACM Multimedia'99, pp: 67-76
9. Liu Z, Huang J, Wang Y (1998) Classification TV programs based on audio information using hidden Markov model. In: IEEE Second Workshop on Multimedia Signal Processing, 1998, pp: 27 -32
10. Wold E, Blum T, Keislar D, Wheaten J (1996) Content-based classification, search, and retrieval of audio. In: IEEE Multimedia, Volume: 3 Issue: 3, Fall 1996, pp: 27 -36
11. Zhu Liu, Qian Huang (2000) Content-based indexing and retrieval-by-example in audio. In: ICME 2000, Volume: 2, pp: 877 -880
12. Beritelli F, Casale S, Russo M (1995) Multilevel Speech Classification Based on Fuzzy Logic. In: Proceedings of IEEE Workshop on Speech Coding for Telecommunications, 1995, pp: 97-98
13. Zhu Liu, Qian Huang (1998) Classification of audio events in broadcast news. In: IEEE Second Workshop on Multimedia Signal Processing, 1998, pp:364 -369
14. Mingchun Liu, Chunru Wan (2001) A study on content-based classification and retrieval of audio database. In: International Database Engineering and Application Symposium, 2001, pp: 339-345
15. Li S.Z (2000) Content-based audio classification and retrieval using the nearest feature line method, IEEE Transactions on Speech and Audio Processing, Volume: 8 Issue: 5, Sept 2000, pp: 619 -625
16. Jang J.-S.R (1993) ANFIS: adaptive-network-based fuzzy inference system, IEEE Transactions on Systems, Man and Cybernetics, 1993, volume: 23, Issue: 3, pp: 665-685

Soft Computing Technology for Dynamic Web Pages Categorization

Vincenzo Loia

Dipartimento di Matematica ed Informatica
Università di Salerno
84081 Baronissi (Salerno), Italy, loia@unisa.it

Summary. Catalogues play an important role in most of the current Web search engines. The catalogues, which organize documents into hierarchical collections, are maintained manually increasing difficulty and costs due to the incessant growing of the WWW. This problem has stimulated many researches to work on automatic categorization of Web documents. In reality, most of these approaches works well either on special types of documents or on restricted set of documents. This paper presents an evolutionary approach useful to construct automatically the catalogue as well as to perform the classification of a Web document. This functionality relies on a genetic-based fuzzy clustering methodology that applies the clustering on the context of the document, as opposite to content-based clustering that works on the complete document information.

1 Introduction

The World Wide Web (WWW or Web) is a cheap and powerful environment for sharing information among specialized communities. The unexpected widespread use of the WWW, the presence of heterogeneous data sources, the absence of recognized organization models, make difficult, in many cases frustanting, the task of Internet searching. One solution to this problem is to categorize the Web documents according to their topics. This explains why popular engines (Altavista, Netscape and Lycos) changed themselves from crawler-based into a Yahoo!-like directories of web sites. Just to give an example of the difficulty of this task, Yahoo! maintains the largest directory list composed of 1.2 million of terms thanks to the support of thousands of human editors.

Many researches have been involved in the study of automatic categorization. Good results have been reported in case of categorization of specific documents, such as newspapers [7] and patent documents [14]. Infoseek experimented neural network technology, other approaches have used clusters generated in a dynamic mode [17] [8].

The impressive evolution of the Web makes difficult the management of consistent category directories. This drawback has an immediate effect in a lost of precision reported by the most popular Web search engines (they return only a fraction of the URLs of interest to user [18], have a small coverage of available data [11], suffer of instability in output for same queries submissions [19].

This work presents a fuzzy clustering approach to Web document categorization [12] [13]. Our approach enable to face with with positive results, the two fundamental problems of Web clustering: the high dimensionality of the feature space and the knowledge of the entire document. The first problem is tackled with an evolutionary approach. The genetic computation assures stability and efficiency also in presence of a large amount of data. About the second issue we perform a clustering based on the analysis of the context rather than the content of the document. Context-based fuzzy clustering strongly reduces the size of the Web document to process, without grave fall of performances.

2 A Contextual View of a Web Page

Let us consider a link in a Web page, as shown in Figure 1: in general we note the existence of sufficient information spent to describe the referenced page. Thus this information may be used to categorize a document. The process starts with an initial list of URLs, and, for each URL, retrieves the web document, analyzing the structure of the document expressed in terms of its HTML tags. For each meaningful tag, contextual data are extracted. For example, when the <A> tag is found containing an URL, an URL Context Path (URL: C_1: C_2:...: C_n) is defined, containing the list of the context strings C_i so far associated to the URL. For example, let us consider the following fragment of an HTML page from Altavista:

The following context paths are created:

1. "http://www.dc.turkuamk.fi/LDP/LDP/nag/node27.html"
   "IP Routing"
       "Next: IP Networks Up: Issues of TCP/IP Networking Previous: Address Resolution. IP Routing. IP Networks. Subnetworks. Gateways. The Routing Table...."
   "Networking Gateways"

2. "http://pyx.net/":
   "Pyx Networking Solutions. "
       "Welcome to Pyx Networking Solutions. Since 1973, Pyx has been providing quality computer services at great prices to a variety of clients, including..."

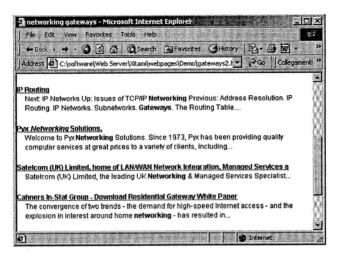

Fig. 1. Example of contexts in a Web page.

"Networking Gateways"

3. "http://www.satelcom.co.uk/":
   "Satelcom (UK) Limited, home of LAN/WAN Network Integration, Managed Ser-
   vices a "
   "Satelcom (UK) Limited, the leading UK Networking & Managed Ser-
   vices Specialist..."
   "Networking Gateways"

4. "http://www.instat.com/catalog/downloads/resgateway.asp":
   "Cahners In-Stat Group - Download Residential Gateway White Paper "
   "The convergence of two trends - the demand for high-speed Internet
   access - and the explosion in interest around home networking - has
   resulted in..."
   "Networking Gateways"

Any URL is analyzed through a breadth-first visiting: first the complete
page is analyzed, then for each external link a new visiting is triggered on
the corresponding host. Next step regards the clustering process that exploits
the Context Paths database and the categories-based catalogue in order to
evaluate the membership value of each URL to a set of categories.

3 Architecture

Usually a Web search engine exploits two basic technologies for document retrieval: - *indexing* the Web page is indexed by a number of words or phrases
representing an abbreviated version of the page itself; - *directories* the page
is represented by a position within a knowledge hierarchy. This section shows
how our system enables to classify a Web document with a precision comparable with a directory approach and with a dimensionality and updating speed
comparable with an indexing technique. Our system returns a database of the

most meaningful categories that characterize a Web area (a set of URLs) under analysis. This task is done thanks to an evolutionary process that updates the previous, existing catalogue. Figure 2 shows the overall architecture, by detailing the role of each basic module.

At instant t_0 we assume the availability of an initial catalogue, used as a kind of training set. The evolved catalogue, containing new category entries, is then used to classify the Web documents. The system is based on a client-server architecture in order to distribute the computational agents charged to load the document from the Web and to classify the document itself.

The evolution layer consists of different modules: (1) on the client-side the *SpiderAgents* have been implemented in order to acquire the context paths of the Web documents, (2) on the server-side the software agents *Genetic Engine* have been realized in order to collect the context paths and to transform them into genotypes. This enables to produce, through the genetic-based process, the catalogue, and (3) the agents *Clusterizer* has been designed to classify the Web documents.

Here follows a short discussion about the basic technologies employed for the automatic categorization.

spidering: the goal of the spidering process is to perform a parsing of the document in order to extract the information concerning the context paths;

classification: we use a model of *context fuzzy clustering*, based on syntax analysis (part of speech) and semantic analysis (WordNet [21]) of the information derived from the context paths;

evolution of the category catalogue: the context fuzzy clustering is embedded into a genetic framework able to produce automatically an updating procedure on the catalogue.

The system is written in Java 2 [9], the distributed computation is managed using **Remote Method Invocation (RMI)** technology supported by the SUN platform **JDK**.

4 Clustering Methodology

Let **T** be the set of the noun phrases. $\forall x \in T$ we define $\tilde{x}$ as the *fuzzyset* "noun phrases *similar to* x", formally:

$$\tilde{x} = \{(t, \mu_x(t)) \mid \forall t \in T\}$$

with $\mu_x : T \to [0, 1]$ as membership function.

The function is defined in order to give higher values for the noun phrase that generalize the original term of the category. Its calculus takes into account the synonyms for each simple term contained into the noun phrase of the category, rejecting the terms that are not synonyms or related terms.

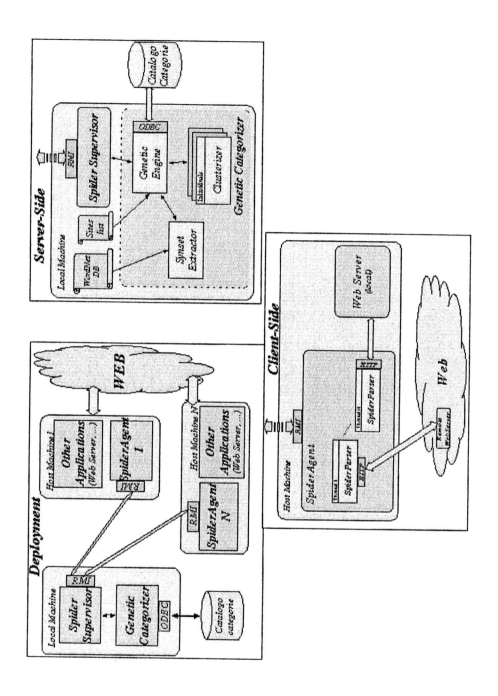

Fig. 2. Software Architecture - Evolution phase.

Any synonym of the simple term has a weight: the weights are higher for hypernym synonyms (generalization terms) and lower for hyponym synonyms (specialization terms), hence the clustering method brings up generalization with respect to each document matched. The membership value of a noun phrase, derived from a combination of simple terms, is given as an average of the synonyms weights.

Given $P(T)$ as the power set of T, let us define the following similarity measure:

Let $x = (t_1, \ldots, t_n) \in P(T)$ and $t_i \in T$ $\forall i = 1..n$
$y = (h_1, \ldots, h_p) \in P(T)$ and $h_j \in T$ $\forall j = 1..p$

$$S_K(x, y) = \sum_{j=1}^{p} \sum_{i=1}^{n} (\mu_{t_i}(h_j))^K \qquad \text{(shortly } x \oplus_k y) \qquad (1)$$

where K is the *similarity factor* of the measure.

Given a couple $(x, y) \in P(T)^2$ we define $G: P(T) \times P(T) \rightarrow [0, 1]$ as the *coverage* of y on x:

$$G(x, y) = \frac{|\{h_j| \, h_j \in y \text{ and } \exists \, t_i \in x \ni' \mu_{t_i}(h_j) > 0\}|}{|x|} \qquad \text{(shortly } x \sqcap y)$$
$$(2)$$

Each category (or sub-category), defined by its noun phrases, is viewed as a cluster $C_j \in P(T)$. Objects of the cluster are URLs extracted from the Web documents: each URL has an associated **Context Path** as *feature vector*, represented by $CP_i \in P(T)$ (for the ith context path).

In order to evaluate the membership grade μ_{ij} of the CP_i on cluster C_j, a *familiarity grade* A_{ij} is defined; this parameter is the weight returned by the matching between context path and category, computed as the similarity measure on $P(T)$ between C_j and CP_i.

Up now the clusters are statically defined (their noun phrases are fixed). The dynamical behavior is provided by the genetic exploration (as defined in the next paragraph) and by a *specialization grade* s for each cluster, that allows us to vary the cluster dimension. The specialization grade exploits the *similarity factor K* that enables to modify the incidence of each similarity grade for the single terms. The next formula defines the familiarity grade using the specialization grade s_j for cluster C_j.

Familiarity Grade:

$$A_{ij} = \frac{C_j \oplus_{s_j} CP_i}{\textit{noun phrases matched by } CP_i \textit{ on } C_j} \qquad (3)$$

$A_{ij} \in [0, 1]$

Membership Grade:

$$\mu_{ij} = A_{ij} \cdot (C_j \sqcap CP_i) \quad \mu_{ij} \in [0, 1] \qquad (4)$$

Our clustering method exploits the concept of the *overlapping* flexibility; it allows objects to belong to all clusters.

Overlapping Property:

$$\sum_{j=1}^{|C|} \mu_{ij} \geq 0 \tag{5}$$

Finally, the clustering method maximizes the following *Index of Quality* $J(C)$, for which an *Influence Grade* **m** is introduced in order to reduce the impact of lower μ_{ij} values. At the increasing of **m** more relevant will be the weight of the clusters characterized by a higher specialization (membership grade).

Index of Quality:

$$J(C) = \sum_{j=1}^{C}(J_j) \tag{6}$$

$$J_j = \begin{cases} (\sum_{i=1}^{N} \mu_{ij})^m & \text{no subcategory in } C_j \\ ((\sum_{i=1}^{N} \mu_{ij} + 1) \cdot \overset{subcategs \ C_j}{\underset{c}{\sum}} J_c)^m & \text{otherwise} \end{cases} \tag{7}$$

with $m \in [1,\infty)$ and J_j as Index of Quality for the jth category.

Index of Quality is skilled to specialize the categories, in order to contrast the generalization spur arising from the computation of matching weights.

5 Genetic Framework

1. **Representation of genomes** – the genome is constituted by a representation of a structured hierarchy of thematic categories, named *Category Forest*, composed by a set of tree structures, our *Category Tree*. Each Category Tree is viewed as a Root Category (it identifies a thematic category). Starting from a Root Category we find the subcategory nodes (specialization of category) which may be parents of other subcategories of lower level as shown in Figure 3.
   Each root node is supported by three threshold values useful for the specialization grade of the thematic category (each subcategory is accompanied by a specialization grade).
   The subcategories can be defined **fixed** in the parent category, by means of a marker; this is useful to do not move the subcategory into other parent categories as effect of the mutation operator.
   Figure 4 shows the information contained into each type of node.
2. **Definition of the fitness function** - It is composed of two different evaluations. The first, named *Clustering Fitness*, is computed by the clustering methodology in terms of Index of Quality. The second factor is the

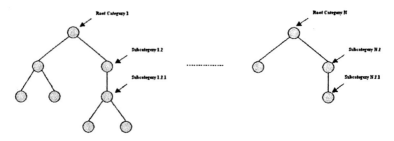

Fig. 3. Graphical representation of category concept.

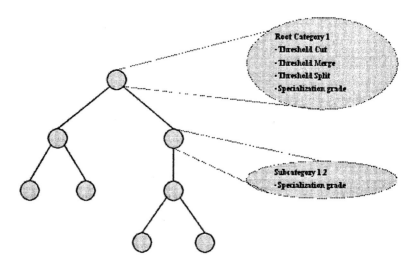

Fig. 4. Fundamental information inside a node.

Quality of Distribution (QoD), measuring the quality of distribution of the documents into thematic categories. This value is computed by averaging the membership grades of the documents, for each category or subcategory.

- **Clustering Fitness (Index of Quality):**

$$J(C) = \sum_{j=1}^{C} (J_j)$$

164

$$J_j = \begin{cases} (\sum\limits_{i=1}^{N} \mu_{ij})^m & \text{no subcategory in } C_j \\ ((\sum\limits_{i=1}^{N} \mu_{ij} + 1) \cdot \overset{subcategories\ C_j}{\underset{c}{\sum}} J_c)^m & \text{otherwise} \end{cases}$$

- **Quality of Distribution (QoD):**

$$QoD = \frac{\sum(QoD_{category})}{\#root\ categories}$$

$$QoD_{category} = \frac{\widehat{\mu} + \sum(QoD_{category}\ of\ the\ subcategories)}{\#subcategories + 1}$$

where $\widehat{\mu}$ is the average of membership values of the documents into the category (root category or subcategory).
- **Fitness function of the individual:**

$$Fitness = QoD * ClusteringFitness$$

3. **Definition of the Crossover operator** – The crossover point is chosen randomly taking into account that root categories that can not be broken by crossover (see Figure 5).

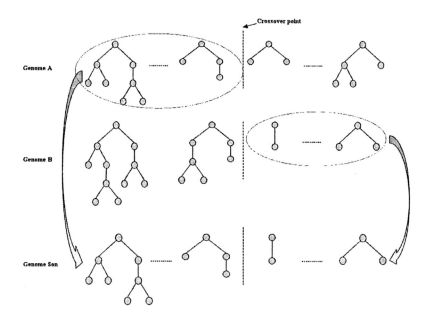

Fig. 5. Crossover effect on categories.

4. **Definition of mutation operators** – The following mutation operators are defined:

- **Mutation Cutting** – Choose randomly both a root category and a subcategory into it: the subcategory is removed with its subtree (see Figure 6).

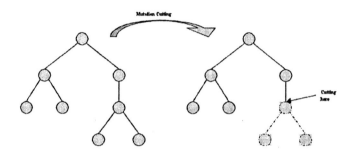

Fig. 6. Mutation Operator : Cutting.

- **Mutation Merging** – Choose randomly a root category and extract randomly two "child" subcategories (same parent category). The operator merges the root nodes of the two selected subcategories (see Figure 7).

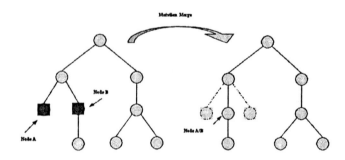

Fig. 7. Mutation Operator : Merging.

- **Mutation Specialization Grade**– Choose randomly a root category and modify its specialization grade.
- **Mutation Exchange Parent (Swap)** – Choose randomly a root category and extract randomly two subcategories with different parent

categories. Hence, the operator swaps the parent categories (see Figure 8).

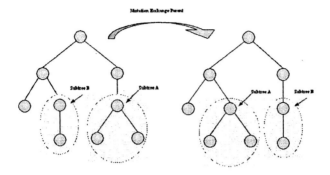

Fig. 8. Mutation Operator : Swap.

Mutation Change Parent – Choose randomly both a root category and a subcategory. Hence, the operator moves (randomly) the subtree in another parent category.

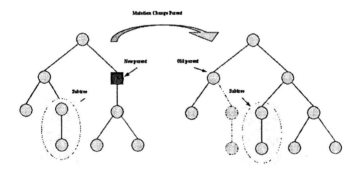

Fig. 9. Mutation Operator : Change Parent.

Stopping criteria

In precedence, we have described as each **generation** works. Genetic algorithm terminates own execution after it has completed a max number of generations, the result will be the genome of the individual with upper fitness value. From this genome, we extract the new categories catalogue evolved.

6 Testing

In order to verify the efficiency of our clustering methodology we take as target the Open Directory Project(ODP) [16] a well known (public domain) project of human categorization of Web documents. We use the synonyms and related terms, computed in advance for each category of the catalogue, using WordNet [21].

Our experiment has been conducted on the following subset of the categories catalogue of ODP :

Science	Health	Arts	Bookmarks
Business	Test	Home	Sports
Private	World	Computers	Regional
Reference	Shopping	Games	News
Society	Recreation		

The URLs, with their short description, are collected in an HTML document in order to extract the corresponding Context Paths.

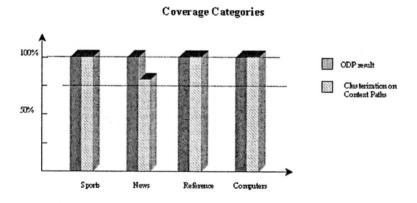

Fig. 10. Coverage of Categories.

Figure 10 reports the behavior of our approach compared with ODP. We obtained automatically the same "human" categorization for the categories Sports, Reference and Computer.

As shown in the figure, the "News" category is not totally covered. This happens because into this category there are URLs not completely described. Below we give an example of context paths of some URLs (contained into ODP database) that our clustering is not able to associate to the right "News" category.

"http://www.bcity.com/bollettino: "

  "International Bulletin"

    "International politics. Italian, French and (some) English."

"http://www.pressdigest.org/ ":

  "Pressdigest"

    "International and Multilingual press digest."

The reason of this drawback is due to the WordNet database: the term "news" is not related to "bulletin" and "digest" as synonyms.

In order to highlight the role of fuzziness, Figure 11 shows the membership value of the URL http://attention.hypermart.net associated to the category "News".

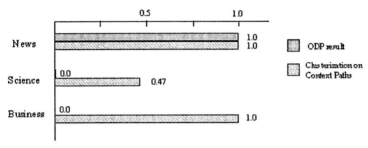

Fig. 11. Membership value.

As noted in Figure 11, the URL is not exclusively associated to the category (as opposite of ODP): this means that in our case, if the user searches URLs about technology into the category "Science" , the search engine shall be able to return a reference to the URL attention.hypermart.net, even though with a membership value lower than the News category.

7 Related Works

The role of cluster, as useful strategy to improve Web search engine behaviors, has reported an increasing interest in these recent years. A well explored issue is to cluster the results of a Web search to better formulate the query. In [4] the query refinement, obtained also thanks to the user's feedback, guarantees a customization of a search space that better fits the user's need. In [2] it is proved how a graph partitioning based clustering technique, without the constraint to specify pre-specified ad-hoc distance functions, can effectively discover Web document similarities and associations. A linear time algorithm which creates clusters on the analysis of phrases shared between Web documents is discussed in [20]. A machine learning approach has been used in [15] and [6] for efficient topic-directed spidering and relevant topic extraction. A fuzzy matching for information retrieval searching is discussed in [5].

About the use of contextual information, the ARC system [3] automatically compiles a list of authoritative Web resources on a topic. In [1] and [10] a focused analysis on the document structure is performed in order to extract concepts useful to build classified directories. These interesting approaches do not support fuzzy partitioning and the search of the better partitioning could suffer of the usual drawbacks concerning crips techniques.

Conclusions

In this paper, we present a methodology able to cluster web document into thematic categories. The clustering algorithm is based on a fuzzy clustering method that searches the best categories catalogue for web document categorization. The categorization is performed by context, this means that the clustering is guided by the context surrounding a link in an HTML document in order to extract useful information for categorizing the document it refer to. This approach enables to be media independent, hence to perform the same strategy also for images, audio and video. As key issue of our clustering methodology we use an evolutionary approach inheriting the benefits of a genetic-level explorations. The positive benchmarks reported by comparing our results with a public-domain, significant category-based catalogue stimulates further development of our research.

References

1. Attardi, G., Di Marco S., and Salvi, D. (1998). Categorisation by Context. *Journal of Universal Compouter Science*, 4:719-736.
2. Boley, D., Gini., M., Gross, R., Hang, E-H., Hasting, K., Karypis, G., Kumar, V., Mobasher, B., and Moore, J. (1999). Partioning-based clustering for Web document categorization *Decision Support System*, 27 (1999) 329-341.
3. Chakrabarti, S., Dom, B., Gibson, D., Kleinberg, J., Rahavan, P., and Rajagopalan, S.(1998). Automatic resource list compilation by analyzing hyperlink structure and associated text. *Seventh International World Wide Web Conference*, 1998.
4. Chang, C-H., and Hsu, C-C. (1997). Customizable Multi-Engine Search tool with Clustering. *Sixth International World Wide Web Conference*, April 7-11, 1997 Santa Clara, California, USA.
5. Cohen, W. (1998). A web-based information system that reasons with structured collections of text. *Agents'98*, 1998.
6. Craven, M., DiPasquo, D., Freitag, D., McCallum, A., Mitchell, T., Nigam, K., and Slattery, S. (1998). Learning to extract symbolic knowledge from the World Wide Web. *AAAI-98*, 1998.
7. Hayes, J., and Weinstein, S. P. (1990). CONSTRUE-TIS: A system for content-based indexing of a database of news stories. *Second Annual Conference on Innovative Applications of Artificial Intelligence*, 1-5.

8. Iwayama, M. (1995). Cluster-based text categorization : a comparison of category search strategies. *SIGIR-95*, pp. 273-280.
9. JDK Java 2 Sun. http://java.sun.com
10. Kruschwitz, U. (2001). Exploiting Structure for Intelligent Web Search. *2001 IEEE International Confernce on System Science*, January 3-6, 2001, Hawaii, IEEE Press.
11. Lawrence, S. and Giles, C. L. (1999). *Nature*, 400:107-109. *Sixteenth International Joint Conference on Artificial Intelligence (IJCAI-99)*.
12. Loia, V. and Luongo, P. (2001). Genetic-based Fuzzy Clustering for Automatic Web Document Categorization, *2001 ACM Symposium Applied Computation*, March 11-14 2001, Las Vegas, USA, ACM Press.
13. Loia, V. and Luongo, P. (2001). An Evolutionary Approach to Automatic Web Page Categorization and Updating, *2001 International Conference on Web Intelligence*, October 23-26, 2001, Maebashi City, Japan.
14. Mase, H., Tsuji, H., Kinukawa, H., Hosoya, Y., Koutani, K., and Kiyota, K. (1996). Experimental simulation for automatic patent categorization. *Advances in Production Management Systems*, 377-382.
15. McCallum, A., Nigam, K., Rennie, J., and Seymore, K. (1999). A Machine Learning Approach to Building Domain-Specific Search Engine. *Sixteenth International Joint Conference on Artificial Intelligence (IJCAI-99)*.
16. Open Directory Project. URL: http://dmoz.org/about.html
17. Sahami, M., Yusufali, S., and Baldoando, M. Q., W. (1998) SONIA: A service for organizing networked information autonomously. *Third ACM Conference on Digital Libraries*.
18. Selberg, E. (1999) *Towards Comprehensive Web Search*. PhD thesis, University of Washington.
19. Selberg,E and Etzioni, O. (2000). On the Instability of Web Search Engine. *RIAO 2000*.
20. Zamir, O., and Etzioni, O. (1988). Web Document Clustering: A Feasibility Demonstration. *SIGIR'98*, Melbourne, Australia, ACM Press.
21. A Lexical Database for English. URL: http://www.cogsci.princeton.edu/ wn/

Text Mining using Fuzzy Association Rules

M.J. Martín-Bautista, D. Sánchez, J.M. Serrano, and M.A. Vila

Dept. of Computer Science and Artificial Intelligence. University of Granada.
C/ Periodista Daniel Saucedo Aranda s/n, 18071, Granada, Spain.
mbautis@decsai.ugr.es

Summary. In this paper, fuzzy association rules are used in a text framework. Text transactions are defined based on the concept of fuzzy association rules considering each attribute as a term of a collection. The purpose of the use of text mining technologies presented in this paper is to assist users to find relevant information. The system helps the user to formulate queries by including related terms to the query using fuzzy association rules. The list of possible candidate terms extracted from the rules can be added automatically to the original query or can be shown to the user who selects the most relevant for her/his preferences in a semi-automatic process.

1 Introduction

The data in the Internet is not organized in a consistent way due to a lack of an authority that supervises the adding of data to the web. Even inside each web site, there is a lack of structure in the documents. Although the use of hypertext would help us to give some homogeneous structure to the documents in the web, and therefore, to use data mining techniques for structure data, as it happens in relational databases, the reality is that nobody follows a unique format to write documents for the web. This represents a disadvantage when techniques such as data mining are applied. This leads us to use techniques specifically for text, as if we were not dealing with web documents, but with text in general, since all of them have an unstructured form.

This lack of homogeneity in the web makes the search process of information in the web by querying not so successful as navigators expect. This fact is due to two basic reasons: first, because the user is not able to represent her/his needs in query terms and second, because the answer set of documents is so huge that the user feels overwhelmed. In this work, we address the first problem of query specification.

Data mining techniques has been broadly applied to text, generating what is called Text Mining. Sometimes, the data mining applications requires the user to know how to manage the tool. In this paper, the rules extracted from texts are not shown to the user specifically. The generated rules are applied to help user to refine the query but the user only see, considering a process non automatic completely, a list of candidate terms to add to the query.

When a user try to express her/his needs in a query, the terms that finally appear in the query are usually not very specific due to the lack of background knowledge of the user about the topic or just because in the moment of the query, the terms do not come to the user's mind. To help the user with the query construction, terms related to the words of a first query may be added to the query. From a first set of documents retrieved, data mining techniques are applied in order to find association rules among the terms in the set. The most accurate rules that include the original query words in the antecedent / consequent of the rule, are used to modify the query by automatically adding these terms to the query or, by showing to the user the related terms in those rules, so the modification of the query depends on the user's decision. A generalization or specification of the query will occur when the terms used to reformulate the query appear in the consequent / antecedent of the rule, respectively. This suggestion of terms helps the user to reduce the set of documents, leading the search through the desired direction.

This paper is organized as follows: in section 2, a summary of literature with the same purpose of this work is included. From section 3 to section 6, general theory about data mining and new proposals in the fuzzy framework are presented. Concretely, in section 3 and 4, the concepts of association rules, fuzzy association rules and fuzzy transactions are presented. In section 5, new measures for importance and accuracy of association rules are proposed. An algorithm to generate fuzzy association rules is presented in section 6. An application of this theory to text framework is proposed in section 7 and 8. The definition of text transactions is given in section 7, while the extracted text association rules are applied to query reformulation in an Information Retrieval framework in section 8. Finally, concluding remarks and future trends are given in section 9.

2 Related Work

One of the possible applications of Text Mining is the problem of query refinement, which has been treated from several frameworks. On the one hand, in the field of Information Retrieval, the problem has been defined as query expansion, and we can find several references with solutions to this problem. A good review in the topic can be found in [20]. On the other hand, techniques such as Data Mining, that have been applied successfully in the last decade in the field of Databases, have been also applied to solve some classical Information Retrieval problems such as document classification [33] and query optimization [46]. In this section, prior work in both frameworks, Information

Retrieval and Data Mining is presented, although the number of approaches presented in the first one, is much more extended than in the second one.

2.1 Previous Research in the Data Mining and Knowledge Discovery Framework

In general terms, the application of Data Mining and Knowledge Discovery techniques to text has been called Text Mining and Knowledge Discovery in Texts, respectively. The main difference to apply these techniques in a text framework is the special characteristics of text as unstructured data, totally different from databases, where mining techniques are usually applied and structured data is managed. Some general approaches about Text Mining and Knowledge Discovery in Texts can be found in [17], [21], [28],[31]

In this work, association rules applying techniques form data mining will be discovered as a process to select the terms to be added to the original query. Some other approaches can be found in this direction. In [46] a vocabulary generated by the association rules is used to improve the query. In [22] a system for Finding Associations in Collections of Text (FACT) is presented. The system takes background knowledge to show the user a simple graphical interface providing a query language with well-defined semantics for the discovery actions based on term taxonomy at different granularity levels. A different application of association rules but in the Information Retrieval framework can be found in [33] where the extracted rules are employed for document classification.

2.2 Previous Research in the Information Retrieval Framework

Several classifications can be made in this field according to the documents considered to expand the query, the selection of the terms to include in the query, and the way to include them. In [50] the authors make a study of expansion techniques based on the set of documents considered to analyze for the query expansion. If these documents are the corpus as a whole, from which all the queries are realized, then the technique is called *global analysis*. However, if the expansion of the query is performed based on the documents retrieved from the first query, the technique is denominated *local analysis*, and the set of documents is called *local set*. This local technique can also be classified into two types. On the one hand, *local feedback* adds common words from the top-ranked documents of the local set. These words are identified sometimes by clustering the document collection [3]. In this group we can include the relevance feedback process, since the user have to evaluate the top ranked documents from which the terms to be added to the query are selected. On the other hand, *local context analysis* [50], which combines global analysis and context local feedback to add words based on relationships of the top-ranked documents. The co-occurrences of terms are calculated based on passages (text windows of fixed size), as in global analysis, instead of complete documents. The authors show that, in general, local analysis performs better than global one.

In our approach, both a global and a local technique are considered. On the one hand, association rules will be extracted from the corpus and applied to expand the query, and on the other hand, only the top ranked documents will be considered to carry out the same process.

Regarding the selection of the terms, some approaches use several techniques to identify terms that should be added to the original query. The first group is based on their association relation by co-occurrence to query terms [47]. Instead of simply terms, in [50] find co-occurrences of concepts given by noun groups with the query terms. Some other approaches based on concept space are [12]. The statistical information can be extracted from a clustering process and ranking of documents from the local set, as it is shown in [13] or by similarity of the top-ranked documents [36]. All these approaches where a co-occurrence calculus is performed has been said to be suitable for construct specific knowledge base domains, since the terms are related, but it can not be distinguished how [8]. The second group searches terms based on their similarity to the query terms, constructing a similarity term thesaurus [41]. Other approaches in this same group, use techniques to find out the most discriminatory terms, which are the candidates to be added to the query. These two characteristics can be combined by first calculating the nearest neighbors and second by measuring the discriminatory abilities of the terms [38]. The last group is formed by approaches based on lexical variants of query terms extracted from a lexical knowledge base such as Wordnet [35]. Some approaches in this group are [49], and [8] where a semantic network with term hierarchies is constructed. The authors reveal the adequacy of this approach for general knowledge base, which can be identified in general terms with global analysis, since the set of documents from which the hierarchies are constructed is the corpus, and not the local set of a first query. Previous approaches with the idea of hierarchical thesaurus can be also found in the literature, where an expert system of rules interprets the user's queries and controls the search process [25].

In our approach, since we are performing a local analysis, fuzzy association rules are used as a technique to find relations among the terms. The aim of the use of this technique is detail and give more information by means of inclusion relations about the connection of the terms, avoiding the inherent statistical nature of systems using co-occurrences as relationships among terms, which performance is only good where the terms selected to expand the query comes from relevant documents of the local set [27]. Previous good results of the use of fuzzy association rules in comparison with crisp association rules and pure statistical methods have been presented in the relational database framework [4], [16], [18].

As for the way to include the terms in the query, we can distinguish between automatic and semi-automatic query expansion [41]. In the first group, the selected terms can substitute or be added to the original query without the intervention of the user [10], [25], [47]. In the second group, a list of candidate terms is shown to the user, which makes the selection [48]. Generally, automatic query expansion is used in local analysis and semi-automatic query expansion is more adequate for global analysis, since the user has to decide from a broad set of terms from the corpus which are more related to her/his needs.

3 Association Rules

The obtaining and mining of association rules is one of the main research problems in data mining framework [1]. Given a database of transactions, where each transaction is an itemset, the obtaining of association rules is a process guided by the constrains of *support* and *confidence* specified by the user. Support is the percentage of transactions containing an itemset, calculated in a statistical manner, while confidence measures the strength of the rule. Formally, let T be a set of transactions containing items of a set of items I. Let us consider two itemsets $I_1, I_2 \subseteq I$, where $I_1 \cap I_2 = \emptyset$. A rule $I_1 \Rightarrow I_2$ is an implication rule meaning that the apparition of itemset I_1 implies the apparition of itemset I_2 in the set of transactions T. I_1 and I_2 are called antecedent and consequent of the rule, respectively. Given a support of an itemset noted by $supp(I_k)$, and the rule $I_1 \Rightarrow I_2$, the support and the confidence of the rule noted by $Supp(I_1 \Rightarrow I_2)$ and $Conf(I_1 \Rightarrow I_2)$, respectively, are calculated as follows:

$$Supp(I_1 \Rightarrow I_2) = supp(I_1 \cup I_2) \qquad (1)$$

$$Conf(I_1 \Rightarrow I_2) = \frac{supp(I_1 \cup I_2)}{supp(I_1)} \qquad (2)$$

The constrains of minimum support and minimum confidence are established by the user with two threshold values: *minsupp* for the support and *minconf* for the confidence. A *strong rule* is an association rule whose support and confidence are greater that thresholds minsupp and minconf, respectively. Once the user has determined these values, the process of obtaining association rules can be decomposed in two different steps:

Step 1.- Find all the itemsets that have a support above threshold *minsupp*. These itemsets are called *frequent itemsets*.
Step 2.- Generate the rules, discarding those rules below threshold *minconf.*

The rules obtained with this process are called boolean association rules in the sense that they are generated from a set of boolean transactions where the values of the tuples are 1 or 0 meaning that the attribute is present in the transaction or not, respectively.

The application of these processes is becoming quite valuable to extract knowledge in business world. This is the reason why the examples given in the literature to explain generation and mining processes of association rules are based, generally, on sale examples of customers shopping. One of the most famous examples of this kind is the market basket example introduced in [1], where the basket of customers is analyzed with the purpose of know the relation among the products that everybody buy usually. For instance, a rule with the form *bread⇒milk* means that everybody that buy bread also buy milk, that is, the products bread an milk usually appears together in the market basket of customers. We have to take into account, however, that this rule obtaining has an inherent statistical nature, and is the role of an expert the interpretation of such rules in order to extract the knowledge that reflects human behavior. This fact implies the generation of easy rules understandable for an expert of the field described by the rules, but probably with no background knowledge of the data mining concepts and techniques.

The consideration of rules coming from real world implies, most of the times, the handling of uncertainty and quantitative association rules, that is, rules with quantitative attributes such as, for example, the age or the weight of a person. Since the origin of these rules is still considered as a set of boolean transactions, a partition into intervals of the quantitative attributes is needed in order to transform the quantitative problem in a boolean one. The discover of suitable intervals with enough support is one of the problems to solve in the field proposed and addressed in several works [14], [23], [39]. In the first work, an algorithm to deal with non binary attributes, considering all the possible values that can take the quantitative attributes to find the rules. In the last two works, however, the authors strengthen the suitability of the theory of fuzzy sets to model quantitative data and, therefore, deal with the problem of quantitative rules. The rules generated using this theory are called fuzzy association rules, and their principal bases as well as the concept of fuzzy transactions are presented in next section.

4 Fuzzy Transactions and Fuzzy Association Rules

Fuzzy association rules are defined as those rules that associate items of the form *(Attribute, Label)*, where the label has an internal representation as fuzzy set over the domain of the attribute [18]. The obtaining of these rules comes from the consideration of fuzzy transactions. In the following, we present the main and features related to fuzzy transactions and fuzzy association rules. The complete model and applications of these concepts can be found in [14].

4.1 Fuzzy Transactions

Given a finite set of items I, we define a fuzzy transaction as any nonempty fuzzy subset $\tilde{\tau} \subseteq I$. For every $i \in I$, the membership degree of i in a fuzzy transaction $\tilde{\tau}$ is noted by $\tilde{\tau}(i)$. Therefore, given an itemset $I_o \subseteq I$, we note $\tilde{\tau}(I_0)$ the membership degree of I_0 to a fuzzy transaction $\tilde{\tau}$. We can deduce from this definition that boolean transactions are a special case of fuzzy transactions. We call FT-set the set of fuzzy transactions, remarking that it is a crisp set.

A set of fuzzy transactions FT-set is represented as a table where columns and rows are labeled with identifiers of items and transactions, respectively. Each cell of a pair *(transaction, itemset)* of the form $(I_0, \tilde{\tau}_j)$ contains the membership degree of I_0 in $\tilde{\tau}_j$, noted $\tilde{\tau}_j(I_0)$ and defined as

$$\tilde{\tau}(I_0) = \min_{i \in I_0} \tilde{\tau}(i) \tag{3}$$

The representation of an item I_0 in a FT-set T based in I is represented by a fuzzy set $\tilde{\Gamma}_{I_0} \subseteq T$, defined as

$$\tilde{\Gamma}_{I_0} = \sum_{\tilde{\tau} \in T} \tilde{\tau}(I_0)/\tilde{\tau} \tag{4}$$

4.2 Fuzzy Association Rules

A fuzzy association rule is a link of the form $A \Rightarrow B$ such that $A, B \subset I$ and $A \cap B = \emptyset$, where A is the antecedent and B is the consequent of the rule, being both of them fuzzy itemsets. An ordinary association rule is a fuzzy association rule. The meaning of a fuzzy association rule is, therefore, analogous to the one of an ordinary association rule, but the set of transactions where the rule holds, which is a FT-set. If we call $\tilde{\Gamma}_A$ and $\tilde{\Gamma}_B$ the degrees of attributes A and B in every transaction $\tilde{\tau} \in T$, we can assert that the rule $A \Rightarrow B$ holds with totally accuracy in T when $\tilde{\Gamma}_A \subseteq \tilde{\Gamma}_B$.

5 Importance and Accuracy Measures for Fuzzy Association Rules

The imprecision latent in fuzzy transactions makes us consider a generalization of classical measures of support and confidence by using approximate reasoning tools. One of these tools is the evaluation of quantified sentences presented in [51]. A quantified sentence is and expression of the form "Q of F are G", where F and G are two fuzzy subsets on a finite set X, and Q is a relative fuzzy quantifier. We focus on quantifiers representing fuzzy percentages with fuzzy values in the interval [0,1] such as "most", "almost all" or "many". These quantifiers are called *relative quantifiers*.

Let us consider Q_M a quantifier defined as $Q_M(x) = x, \forall x \in [0, 1]$. We define the **support of an itemset** I_0 in an FT-set T as the evaluation of the quantified sentence,

$$Q_M \ of \ T \ are \ \tilde{\Gamma}_{I_0} \tag{5}$$

while the **support of a rule** $A \Rightarrow B$ in T is given by the evaluation of

$$Q_M \ of \ T \ are \ \tilde{\Gamma}_{A \cup B} = Q_M \ of \ T \ are \ \tilde{\Gamma}_A \cap \tilde{\Gamma}_B \tag{6}$$

and its confidence is the evaluation of

$$Q_M \ of \ \tilde{\Gamma}_A \ are \ \tilde{\Gamma}_B \tag{7}$$

We evaluate the sentences by means of method GD presented in [19]. To evaluate the sentence "Q of F are G", a compatibility degree between the relative cardinality of G with respect to F and the quantifier is represented by $GD_Q(G/F)$ and defined as

$$GD_Q(G/F) = \sum_{\alpha_i \in \Delta(G/F)} (\alpha_i - \alpha_{i+1}) \cdot Q\left(\frac{\left|(G \cap F)_{\alpha_i}\right|}{|F_{\alpha_i}|} \right) \tag{8}$$

where $\Delta(G/F) = \Lambda(G \cap F) \cup \Lambda(F)$, $\Lambda(F)$ being the level set of F, and $\Delta(G/F) = \{\alpha_1, \ldots, \alpha_p\}$ with $\alpha_i > \alpha_{i+1}$ for every $i \in \{1, \ldots p\}$. The set F is assumed to be normalized. If not, F is normalized and the normalization factor is applied to $G \cap F$.

We must point out, moreover, that when we are dealing with crisp data in a T-set T, the evaluation of sentences are the ordinary measures of support and confidence of crisp association rules. Therefore, the evaluation of sentence "Q of F are G" is

$$Q\left(\frac{|F \cap G|}{|F|} \right) \tag{9}$$

when F and G are crisp. The GD method verifies this property. For more details, see [19]. We can interpret the ordinary measures of confidence and support as the degree to which the confidence and support of an association rule is Q_M. Other properties of this quantifier can be seen in [14].

This generalization of the ordinary measures allow us, using Q_M, provide an accomplishment degree, basically. Hence, for fuzzy association rules we can assert

$$Q_M T \in T, \ A \Rightarrow B \tag{10}$$

5.1 Certainty as a New Measure for Rule Accuracy

We propose the use of certainty factors to measure the accuracy of association rules. A previous study can be found in [15]. Certainty factors were developed as a model for the representation of uncertainty and reasoning in rule-based systems [45], although they have been used in knowledge discovery too [24].

180

We define certainty factor (CF) of a fuzzy association rule $A \Rightarrow B$ based on the value of the confidence of the rule. If $Conf\,(A \Rightarrow B) > supp\,(B)$ the value of the factor is given by expression (11); otherwise, is given by expression (12), considering that if $supp(B)=1$, then $CF\,(A \Rightarrow B) = 1$ and if $supp(B)=0$, then $CF\,(A \Rightarrow B) = -1$

$$CF\,(A \Rightarrow B) = \frac{Conf\,(A \Rightarrow B) - supp\,(B)}{1 - supp\,(B)} \qquad (11)$$

$$CF\,(A \Rightarrow B) = \frac{Conf\,(A \Rightarrow B) - supp\,(B)}{supp\,(B)} \qquad (12)$$

We demonstrated in [7] that certainty factors verify the three properties by [29]. From now on, we shall use certainty factors to measure the accuracy of a fuzzy association rule. We consider a fuzzy association rule as strong when its support and certainty factor are greater than thresholds $minsupp$ and $minCF$, respectively.

6 Generation of Fuzzy Association Rules

Several approaches can be found in the literature where efficient algorithms for association rule generation like Apriori and AprioriTid [2], OCD [34], SETM [30], DHP [37], DIC [9], FP-Growth [26] and TBAR [6], have been presented. Most of them include and describe the process of generating fuzzy association rules with two basic steps, as we mentioned in Sect. 3: the generation of frequent itemsets and the obtaining of the rules, with their associated grades of support and confidence. As we are considering fuzzy association rules, in Algorithm 1, we show a process to find the frequent itemsets. For this purpose, the transactions are analyzed one by one and the itemsets whose support is greater than threshold minsupp are selected. The items are processed ordered by size. First 1-itemsets, next 2-itemsets and so on. The variable l stores the actual size. The set L_l stores the l-itemsets that are being analyzed and, at the end, it stores the frequent l-itemsets.

In order to deal with fuzzy transactions, we need to store the difference between the cardinality of every α-cut of $\tilde{\Gamma}_{I_0}$ and the cardinality of the corresponding strong α-cut, $\alpha \in [0, 1]$, for all considered itemsets I_0. Specifically

$$\left| \left(\tilde{\Gamma}_{I_0} \right)_\alpha \right| - \left| \left(\tilde{\Gamma}_{I_0} \right)_{\alpha+} \right|$$

where $\left(\tilde{\Gamma}_{I_0} \right)_\alpha = \left\{ \tilde{\tau} \in T \;\middle|\; \tilde{\Gamma}_{I_0}\,(\tilde{\tau}) \geq \alpha \right\}$ and $\left(\tilde{\Gamma}_{I_0} \right)_{\alpha+} = \left\{ \tilde{\tau} \in T \;\middle|\; \tilde{\Gamma}_{I_0}\,(\tilde{\tau}) > \alpha \right\}$

We use a used a fixed number of k equidistant α-cuts, (specifically k=100, although we a lesser value would be sufficient). By this information, we obtain the fuzzy cardinality of the representation of the items, which is stored in an array V_{I_0}. This array can be easily obtained from an FT-set by adding 1 to $V_{I_0}\left(\tilde{\Gamma}_{I_0}(\tilde{\tau})\right)$ for every itemset I_0 each time a transaction $\tilde{\tau}$ is considered. The function $R(x,k)$ maps the real value x to the nearest value in the set of k equidistant levels we are using.

The procedure *CreateLevel(i, L)* generates a set of i-itemsets such that every proper subset with i-1 items is frequent (i.e. is in L_{i-1}). Since every proper subset of a frequent itemset is also a frequent itemset, with this procedure we avoid analyzing itemsets that do not verify this property, saving space and time.

Algorithm 1 Basic algorithm to find frequent itemsets in a FT-set T

Input: a set I of items and an a FT-set T based on I.
Output: a set of frequent itemsets F.

1. {Initialization}
   a) Create an array $V_{\{i\}}$ of size $k+1$ for every $i \in I$
   b) $L_1 \leftarrow \{ \{i\} \mid i \in I \}$
   c) F=0
   d) $l \leftarrow 1$
2. Repeat until $l > |I|$ or $L_l = 0$
   a) For every $\tilde{\tau} \in T$
      i. For every $I_* \in L_l$
         A. $V_{I_*}\left(R\left(\tilde{\Gamma}_{I_*}(\tilde{\tau}),k\right)\right) \leftarrow V_{I_*}\left(R\left(\tilde{\Gamma}_{I_*}(\tilde{\tau}),k\right)\right) + 1$
   b) For every $I_* \in L_l$
      i. Calculate $GD_Q\left(\tilde{\Gamma}_{I_*}/T\right)$
      ii. If $GD_Q\left(\tilde{\Gamma}_{I_*}/T\right) < minsupp \times |T|$
         A. $L_l \leftarrow L_l \backslash \{I_*\}$
         B. Free the memory used by V_{I_*}
   c) {Variables updating}
      i. $F = F \cup L_l$
      ii. $L_{l+1} \leftarrow CreateLevel\left(l+1, L_l\right)$
      iii. $l \leftarrow l + 1$
3. Return(F)

The complexity of this algorithm is an exponential function of the number of items. The hidden constant is increased in a factor that depends on k as this value affects the size of the arrays V. For more details of the algorithm, see [14]

Once we have obtained the frequent itemsets with the former algorithm, we obtain the confidence by calculating $GD_Q(B/A)$ from V_A and $V_{A \cup B}$. From confidence and support of the consequent, both available, we obtain the certainty factor of the rules. Finally, we can identifier the strong rules by analyzing the values of support and certainty for the rules.

7 Text Mining for Information Access

The main problem when the general techniques of data mining are applied to text, is to deal with unstructured data, in comparison to structured data coming from relational databases. Therefore, with the purpose to perform a knowledge discovery process, we need to obtain some kind of structure in the texts. Different representations of text for association rules extraction have been considered: bag of words, indexing keywords, term taxonomy and multi-term text phrases [17]. In our case, we use automatic indexing techniques coming from Information Retrieval [44]. We represent each document by a set of terms with a weight meaning the presence of the term in the document. Some weighting schemes for this purpose can be found in [43]. One of the more successful and more used representation schemes is the *tf-idf* scheme, which takes into account the term frequency and the inverse document frequency, that is, if a term occurs frequently in a document but infrequently in the collection, a high weight will be assigned to that term in the document. This is the scheme we consider in this work. The algorithm to get the representation by terms and weights of a document d_i can be detailed by the known following steps in Algorithm 2.

Algorithm 2 Basic algorithm to obtain the representation of documents in a collection

---

Input: a set of documents $D = \{d_1, \ldots d_n\}$.
Output: a representation for all documents in D.

1. Let $D = \{d_1, \ldots d_n\}$ be a collection of documents
2. Extract an initial set of terms S from each document $d_i \in D$
3. Remove stop words
4. Apply stemming (*via* Porter's algorithm [40])
5. The representation of d_i obtained is a set of keywords $\{t_1, \ldots, t_m\} \in S$ with their associated weights $\{w_1, \ldots, w_m\}$

We must point out that, as it has been commented and shown in [21], [42], standard Text Mining usually deal with categorized documents, in the sense of documents which representation is a set of *keywords*, that is, terms that really describe the content of the document. This means that usually a full text is not considered and its description is not formed by all the words in the document, even without stop words, but also by keywords. The authors justify the use of keywords because of the appearing of useless rules. Some additional commentaries about this problem regarding the poor discriminatory power of frequent terms can be found in [38], where the authors comment the fact that the expanded query may result worst performance than the original one due to the poor discriminatory ability of the added terms.

However, in document collections where the categorization is not always available, full text is necessary to be considered as starting point. Additionally, special pre-processing tasks of term extraction and selection can be applied to get keywords in these collections. We are not referring here to statistical counts of term occurrences and assigning of weighting schemes such as the *tf-idf* one, but to more elaborated methods that imply additional time process, such as term taxonomy construction, thesauri or controlled vocabulary.

Nevertheless, in dynamic environments or systems where the response-time is important, the application of this pre-processing stage may not be suitable. This is the case of the problem we deal with in this work, the query refinement in Internet, where an automatic process would be necessary. Two time constraints have to be into account: first, the fact that not all web documents have identified keywords when is retrieved, or if they have, we do not have the guarantee that the keywords are appropriate in all the cases. Second, in the case of query refinement, information rule must be shown to the user on-line, that is, while she/he is query the system. Therefore, instead of improve document representation in this situation, we can improve the rule obtaining process. The use of alternative measures of importance and accuracy such as the ones presented in Sect. 5 is considered in this work in order to avoid the problem of non appropriate rule generation.

7.1 Text Transactions

From a collection of documents $D = \{d_1, \ldots, d_n\}$ we can obtain a set of terms $I = \{t_1, \ldots, t_m\}$ which is the union of the keywords for all the documents in the collection. The weights associated to these terms are represented by $W = \{w_1, \ldots, w_m\}$. Therefore, for each document d_i, we consider an extended representation where a weight of 0 will be assigned to every term appearing in some of the documents of the collection but not in d_i.

Considering these elements, we can define a *text transaction* $\tau_i \in T$ as the extended representation of document d_i. Without loosing generalization, we can write $T = \{d_1, \ldots, d_n\}$. However, as we are dealing with fuzzy association rules, we will consider a fuzzy representation of the presence of the terms in documents, by using the normalized tf-idf scheme [32]. Analogously to the former case, we can define a set of *fuzzy text transactions* $FT = \{d_1, \ldots, d_n\}$, where each document d_i corresponds to a fuzzy transaction $\tilde{\tau}_i$, and where the weights $W = \{w_1, \ldots, w_m\}$ of the keyword set $I = \{t_1, \ldots, t_m\}$ are fuzzy values.

8 Query Reformulation Procedure

The purpose of this work is to provide a system with a query reformulation ability in order to improve the retrieval process. We represent the query a $Q = \{q_1, \ldots, q_m\}$ with associated weights $P = \{p_1, \ldots, p_m\}$. To obtain a relevance value for each document, the query representation is matched to each document representation, obtained as explained in Algorithm 2. If a document term does not appear in the query, its value will be assumed as 0. The considered operators and measures are the one from the generalized Boolean model with fuzzy logic [11].

The user's initial query generates a set of ranked documents. If the top-ranked documents do not satisfy user's needs, the query improvement process starts. From the retrieved set of documents, association relations are found. As we explain in Sect.2, two different approaches can be considered at this point: an automatic expansion of the query or a semi-automatic expansion, based on the intervention of the user in the selection process of the terms to be added to the query. The complete process in both cases is detailed in the following:

Case 1: Automatic query reformulation process

1. The user queries the system
2. A first set of documents is retrieved
3. From this set, the representation of documents is extracted following Algorithm 2 and fuzzy association rules are generated following Algorithm 1 and the extraction rule procedure.
4. The terms co-occurring in the rules with the query terms are added to the query.
5. With the expanded query, the system is queried again.

Case 2: Semi-automatic query reformulation process

1. The user queries the system
2. A first set of documents is retrieved
3. From this set, the representation of documents is extracted following Algorithm 2 and fuzzy association rules are generated following Algorithm 1 and the extraction rule procedure
4. The terms co-occurring in the rules with the query terms are shown to the user
5. The user selects those terms more related to her/his needs
6. The selected terms are added to the query, which is used to again to query the system

We must point out that, in both cases, the obtained association rules conform a knowledge base specific for the domain of the first query. Where several queries are performed, a broader knowledge base may be constructed, so original queries will be enriched with more terms as the time passes. However, the obtaining of a huge knowledge-based from iterated query expansions even in different domains probably can not be used for any query in a successful way, since additional semantic relation information should be also take into account in order to get a general knowledge-base. As a future proposal, we can think about combine both domain-specific knowledge base and general knowledge base, looking at the terms appearing in association rules together with query terms appear, and searching in a general knowledge-base additional terms, WordNet [35], for instance, with a semantic relation with all the terms in the rule. Some further discussion about this point can be found in [8]

8.1 Generalization and Specialization of a Query

Once the first query is constructed, and the association rules are extracted, we make a selection of rules where the terms of the original query appear. However, the terms of the query can appear in the antecedent or in the consequent of the rule. If a query term appears in the antecedent of a rule, and we consider the terms appearing in the consequent of the rule to expand the query, a generalization of the query will be carried out. Therefore, a generalization of a query gives us a query on the same topic as the original one, but looking for more general information. However, if query term appears in the consequent of the rule, and we reformulate the query by adding the terms appearing in the antecedent of the rule, then a specialization of the query will be performed, and the precision of the system should increase. The specialization of a query looks for more specific information than the original query but in the same topic. In order to obtain as much documents as possible, terms appearing in both sides of the rules can also be considered.

9 Conclusion and Future Work

In this work, an application of traditional data mining techniques in a text framework is proposed. Classical transactions in data mining are first extended to the fuzzy transactions, proposing new measures to measure the accuracy of a rule. Text transactions are defined based on fuzzy transactions, considering that each transaction correspond to a document representation. The set of transactions represents, therefore, a document collection from which the fuzzy association rules are extracted. One of the applications of this process is to solve the problem of refinement of a query, very well known in the field of Information Retrieval. A list of terms extracted from the fuzzy association rules related to the terms in the query can be automatically added to the original query to optimize the search. This process can also be done with the user intervention, selecting the terms more related to her/his preferences.

As future work, we will implement the application of the model to this query reformulation procedure and compare the results with other approaches to query refinement coming from Information Retrieval.

References

1. Agrawal R, Imielinski T, Swami A (1993) Mining Association Rules between Set of Items in Large Databases. Proc. of the 1993 ACM SIGMOD Conference, pp 207-216
2. Agrawal R, Srikant R (1994) Fast algorithms for mining association rules. Proc. Of the 20th VLDB Conference, pp 478-499
3. Attar R, Fraenkel AS (1977) Local Feedback in Full-Text Retrieval Systems. Journal of the Association for Computing Machinery 24(3):397-417
4. Au WH, Chan KCC (1998) An effective algorithm for discovering fuzzy rules in relational databases. Proc. Of IEEE International Conference on Fuzzy Systems, vol II, pp 1314-1319
5. Baeza-Yates R, Ribeiro-Nieto B (1999) Modern Information Retrieval, Addison-Wesley, USA
6. Berzal F, Cubero JC, Marín N, Serrano JM (2001) TBAR: An efficient method for association rule mining in relational databases. Data and Knowledge Engineering 37(1):47-84
7. Berzal F, Blanco I, Sánchez, Vila MA (2002) Measuring the Accuracy and Importance of Association Rules: A New Framework. Intelligent Data Analysis 6:221-235
8. Bodner RC, Song F (1996) Knowledge-Based Approaches to Query Expansion in Information Retrieval. In: McGalla G (ed) Advances in Artificial Intelligence pp 146-158. Springer, New York
9. Brin S, Motwani JD, Ullman JD, Tsur S (1997) Dynamic itemset counting and implication rules for market basket data. SIGMOD Record 26(2):255-264
10. Buckley C, Salton G, Allan J, Singhal A (1993) Automatic Query Expansion using SMART: TREC 3". *Proc. of the 3rd* Text Retrieval Conference. NIST Special Publication 500-225, pp 69-80
11. Buell DA, Kraft DH (1981) Performance Measurement in a Fuzzy Retrieval Environment. Proceedings of the Fourth International Conference on Information Storage and Retrieval, ACM/SIGIR Forum 16(1): 56-62, Oakland, CA
12. Chen H, Ng T, Martinez J, Schatz BR (1997) A Concept Space Approach to Addressing the Vocabulary Problem in Scientific Information Retrieval: An Experiment on the Worm Community System. Journal of the American Society for Information Science 48(1):17-31
13. Croft WB, Thompson RH (1987) I^3R: A New Approach to the Design of Document Retrieval Systems. Journal of the American Society for Information Science 38(6):389-404
14. Delgado M, Marín N, Sánchez D, Vila MA (2001). Fuzzy Association Rules: General Model and Applications. IEEE Transactions of Fuzzy Systems (accepted)
15. Delgado M, Martín-Bautista MJ, Sánchez D, Vila MA (2000). Mining strong approximate dependences from relational databases. Proc. Of IPMU 2000 2:1123-1130. Madrid, Spain
16. Delgado M, Martín-Bautista MJ, Sánchez D, Vila MA (2001) Mining association rules with improved semantics in medical databases. Artificial Intelligence in Medicine 21:241-245
17. Delgado M, Martín-Bautista MJ, Sánchez D, Vila MA (2002) Mining Text Data: Special Features and Patterns. Proc. of EPS Exploratory Workshop on

Pattern Detection and Discovery in Data Mining, pp 140-153. Imperial College Londres, UK

18. Delgado M, Sánchez D, Vila MA (2000) Acquisition of fuzzy association rules from medical data. In Barro S, Marín R (eds) Fuzzy Logic in Medicine. Physica-Verlag

19. Delgado M, Sánchez D, Vila MA (2000) Fuzzy cardinality based evaluation of quantified sentences. International Journal of Approximate Reasoning 23:23-66

20. Efthimiadis E (1996) Query Expansion. Annual Review of Information Systems and Technology 31:121-187

21. Feldman R, Fresko M, Kinar Y, Lindell Y, Liphstat O, Rajman M, Schler Y, Zamir O (1998) Text Mining at the Term Level. Proc. of the 2nd European Symposium of Principles of Data Mining and Knowledge Discovery, pp 65-73

22. Feldman R, Hirsh H (1996) Mining associations in text in the presence of Background Knowledge. Proc. of the Second International Conference on Knowledge Discovery from Databases

23. Fu AW, Wong MH, Sze SC, Wong WC, Wong WL, Yu WK (1998) Finding Fuzzy Sets for the Mining of Fuzzy Association Rules for Numerical Attributes. Proc. of Int. Symp. on Intelligent Data Engineering and Learning (IDEAL'98), pp 263-268, Hong Kong

24. Fu LM, Shortliffe EH (2000) The application of certainty factors to neural computing for rule discovery. IEEE Transactions on Neural Networks 11(3):647-657

25. Gauch S, Smith JB (1993) An Expert System for Automatic Query Reformulation. Journal of the American Society for Information Science 44(3):124-136

26. Han J, Pei J, Yin Y (2000)Mining frequent patterns without candidate generation. Proc. ACM SIGMOD Int. Conf. On Management of Data, pp 1-12. Dallas, TX, USA

27. Harman D (1988) Towards interactive query expansion. Proc. of the Eleventh Annual International ACMSIGIR Conference on Research and Development in Information Retrieval pp 321-331. ACM Press

28. Hearst M (1999) Untangling Text Data Mining. Proc. of the 37th Annual Meeting of the Association for Computational Linguistics (ACL'99). University of Maryland

29. Hearst M (2000) Next Generation Web Search: Setting our Sites. IEEE Data Engineering Bulletin, Special issue on Next Generation Web Search, Gravano L (ed)

30. Houtsma M, Swami A (1995) Set-oriented mining for association rules in relational databases. Proc. Of the 11th International Conference on Data Engineering pp 25-33.

31. Kodratoff Y (1999) Knowledge Discovery in Texts: A Definition, and Applications. In: Ras ZW, Skowron A (eds) Foundation of Intelligent Systems, Lectures Notes on Artificial Intelligence 1609. Springer Verlag

32. Kraft D, Petry FE, Buckles BP, Sadasivan T (1997) Genetic Algorithms for Query Optimization in Information Retrieval: Relevance Feedback. In: Sanchez E, Shibata T, Zadeh LA, (eds) Genetic Algorithms and Fuzzy Logic Systems, Advances in Fuzziness: Applications and Theory 7:157-173, World Scientific

33. Lin SH, Shih CS, Chen MC, Ho JM, Ko MT, Huang YM (1998) Extracting Classificadion Knowledge of Internet Documents with Mining Term Associations: A Semantic Approach. Proc. of ACM/SIGIR'98 pp 241-249. Melbourne, Australia

34. Mannila H, Toivonen H, Verkamo I (1994) Efficient algorithms for discovering association rules. Proc. Of AAAI Workshop on Knowledge Discovery in Databases pp 181-192
35. Miller G (1990) WordNet: An on-line lexical database. International Journal of Lexicography 3(4)
36. Mitra M, Singhal A, Buckley C (1998) Improving Automatic Query Expansion. Proc. Of ACM SIGIR pp 206-214. Melbourne, Australia
37. Park JS, Chen MS, Yu PS (1995) An effective hash based algorithm for mining association rules. SIGMOD Record 24(2):175-186
38. Peat HJ, Willet P (1991) The limitations of term co-occurrence Data for Query Expansion in Document Retrieval Systems. Journal of the American Society for Information Science 42(5):378-383
39. Piatetsky-Shapiro G (1991) Discovery, Analysis, and Presentation of Strong Rules. In: Piatetsky-Shapiro G, Frawley WJ (eds) Knowledge Discovery in Databases, AAAI/MIT Press
40. Porter MF (1980) An algorithm for suffix stripping. Program 14(3):130-137
41. Qui Y, Frei HP (1993) Concept Based Query Expansion. Proc. Of the Sixteenth Annual International ACM-SIGIR'93 Conference on Research and Development in Information Retrieval pp 160-169
42. Rajman M, Besançon R (1997) Text Mining: Natural Language Techniques and Text Mining Applications. Proc. of the 3rd International Conference on Database Semantics (DS-7)Chapam & Hall IFIP Proceedings serie
43. Salton G, Buckley C (1988) Term-weighting approaches in automatic text retrieval. Information Processing and Management 24(5):513-523
44. Salton G, McGill MJ (1983) Introduction to Modern Information Retrieval. McGraw-Hill
45. Shortliffe E, Buchanan B (1975) A model of inexact reasoning in medicine. Mathematical Biosciences 23:351-379
46. Srinivasan P, Ruiz ME, Kraft DH, Chen J (2001) Vocabulary mining for information retrieval: rough sets and fuzzy sets. Information Processing and Management 37:15-38
47. Van Rijsbergen CJ, Harper DJ, Porter MF (1981) The selection of good search terms. Information Processing and Management 17:77-91
48. Vélez B, Weiss R, Sheldon MA, Gifford DK (1997) Fast and Effective Query Refinement. Proc. Of the 20th ACM Conference on Research and Development in Information Retrieval (SIGIR'97). Philadelphia, Pennsylvania
49. Voorhees EM (1994)Query expansion using Lexical-Semantic Relations. ACM SIGIR pp 61-70
50. Xu J, Croft WB (1996) Query Expansion Using Local and Global Document Analysis. Proc. of the Nineteenth Annual International ACM SIGIR Conference on Research and Development in Information Retrieval pp 4-11
51. Zadeh LA (1983) A computational approach to fuzzy quantifiers in natural languages. Computing and Mathematics with Applications 9(1):149-184

BISC Decision Support System: University Admission System

Masoud Nikravesh[1] and Ben Azvine[2]

[1] BISC Program, Computer Sciences Division, EECS Department
University of California, Berkeley, CA 94720, USA
Email: nikravesh@cs.berkeley.edu
Tel: (510) 643-4522
Fax: (510) 642-5775
URL: http://www-bisc.cs.berkeley.edu

[2] BTexact Technologies
Orion Building pp1/12, Adastral Park,
Martlesham, Ipswich IP5 3RE, UK

Abstract: The process of ranking (scoring) has been used to make billions of financing decisions each year serving an industry worth hundreds of billion of dollars. To a lesser extent, ranking has also been used to process hundreds of millions of applications by U.S. Universities resulting in over 15 million college admissions in the year 2000 for a total revenue of over $250 billion. College admissions are expected to reach over 17 million by the year 2010 for total revenue of over $280 billion. In this paper, we will introduce fuzzy query and fuzzy aggregation as an alternative for ranking and predicting the risk for university admissions, which currently utilize an imprecise and subjective process. In addition we will introduce the BISC Decision Support System. The main key features of the BISC Decision Support System for the internet applications are 1) to use intelligently the vast amounts of important data in organizations in an optimum way as a decision support system and 2) To share intelligently and securely company's data internally and with business partners and customers that can be process quickly by end users.

1 Introduction

Consider walking into a car dealer and leaving with an old used car paying a high interest rate of around 15% to 23% and your colleague leaves the dealer with a luxury car paying only a 1.9% interest rate. Consider walking into a real estate agency and finding yourself ineligible for a loan to buy your dream house. Also consider getting denied admission to your college of choice but your classmate gets accepted to the top school in his dream major. Welcome to the world of ranking, which is used both for deciding college admissions and determining credit risk. In the credit rating world, FICO (Fair Isaac Company) either makes you or breaks you, or can at least prevent you from getting the best rate possible (Fair Isaac). Admissions ranking can either grant you a better educational opportunity or stop you from fulfilling your dream.

When you apply for credit, whether it's a new credit card, a car loan, a student loan, or a mortgage, about 40 pieces of information from your credit card report are fed into a model. That model provides a numerical score designed to predict your risk as a borrower. When you apply for university or college admission, more than 20 pieces of information from your application are fed into the model. That model provides a numerical score designed to predict your success rate and risk as a student. In this paper, we will introduce fuzzy query and fuzzy aggregation as an alternative for ranking and predicting risk in areas which currently utilize an imprecise and subjective process.

The areas we will consider includes university admissions (**Table 1**). Fuzzy query and ranking is robust, provides better insight and a bigger picture, contains more intelligence about an underlying pattern in data and is capable of flexible querying and intelligent searching (Nikravesh 2001a). This greater insight makes it easy for users to evaluate the results related to the stated criterion and makes a decision faster with improved confidence. It is also very useful for multiple criteria or when users want to vary each criterion independently with different degrees of confidence or weighting factor (Nikravesh 2001b).

2 Fuzzy Query

In the case of crisp queries, we can make multi-criterion decision and ranking where we use the functions AND and OR to aggregate the predicates. In the extended Boolean model or fuzzy logic, one can interpret the AND as a fuzzy-MIN function and the OR as a fuzzy-MAX function. Fuzzy querying and ranking is a very flexible tool in which linguistic concepts can be used in the queries and ranking in a very natural form. In addition, the selected objects do not need to match the decision criteria exactly, which gives the system a more human-like behavior.

Table 1. Variables, Granulation and Information used to create the University Admission System Model.

% AP : Advanced Placement	% CP: Course pattern
% IBHL : International Bacculaureat Higher Level (IBHL)	% GPAP: Pattern of Grades through time
% HW: Honors and Awards	% SAT II
% GPA: 12th Grade Courses GPA	% SAT I
% CAoSI: Creative Achievement or Sustained Intellectual	% AAaO: Academic Achievement and Outreach
% ClaCV: Contribution to the intellectual and cultural vitality	% DPBaE: Diversity in the Personal Background and Experience
% Leadership	% Motivation
% Concern: Concern for Community and others	% AAA: Achievements; Art or Athletics
% Employment	% IMajor: Interest in the Major

EthnicName = {'American'; 'Chinese'; 'French'; 'Greek'; 'Indian'; 'Trish'; 'Italian'; 'Japanese';
 'Mediterranean '; 'Persian'; 'Spanish'; 'Taiwanese', 'Not Care'};

Residency={'California Resident'; 'US Resident'; 'International', 'NotCare' };
Sex={'Male'; 'Female, 'Not Care' };   *Minority*={'No'; 'Yes'; 'Not Care'};
HW= {'Few'; 'Some'; 'Lot'; 'Not Care'};   *AAA*= {'Kind of Active'; 'Active'; 'Exceptional'; 'Not Care'};
CP={'Less Than Required'; 'Required'; 'Recommended'; 'Above Recommendation' };
Concern={'Kind of Concern'; 'Concern'; 'Very Concern'; 'Enthusiast'};
Motivation={'Kind of Motivated'; 'Motivated'; 'Highly Motivated'; 'Enthusiast'};
IMajor={'Kind of Interested'; 'Interested'; 'Very Interested'; 'Enthusiast'};
AP= {'Very Low'; 'Low'; 'Medium'; 'High'; 'Very High' };   *IBHL*= {'Very Low'; 'Low'; 'Medium'; 'High'; 'Very High' };
SATI={'Very Low'; 'Low'; 'Medium'; 'High'; 'Very High' };   *SATII*={'Very Low'; 'Low'; 'Medium'; 'High'; 'Very High' };
GPA= {'Very Low'; 'Low'; 'Medium'; 'High'; 'Very High' };   *Employment*={'Few'; 'Average'; 'Kind High'; 'High'; 'Lot'};
CAoSAI= {' Low'; 'Kind Low'; 'Average'; 'Kind of High'; 'High'; 'Exceptional' };
AAaO={' Low'; 'Kind Low'; 'Average'; 'Kind of High'; 'High'; 'Exceptional'};
ClaCV={' Low'; 'Kind Low'; 'Average'; 'Kind of High'; 'High'; 'Exceptional'};
DPBaE={' Low Diversity'; 'Kind Low Diversity'; 'Diverse'; 'Kind of High Diversity'; 'High Diversity'; 'Exceptional'};
Leadership={' Low'; 'Kind Low'; 'Average'; 'Kind of High'; 'High'; 'Exceptional'};

3 University Admissions

Hundreds of millions of applications were processed by U.S. universities resulting in more than 15 million enrollments in the year 2000 for a total revenue of over $250 billion. College admissions are expected to reach over 17 million by the year 2010, for total revenue of over $280 billion. In Fall 2000, UC Berkeley was able to admit about 26% of the 33,244 applicants for freshman admission (University of California-Berkeley). In Fall 2000, Stanford University was only able to offer admission to 1168 men from 9571 applications (768 admitted) and 1257 women from 8792 applications (830 admitted), a general admit rate of 13% (Stanford University Admission).

The UC Berkeley campus admits its freshman class on the basis of an assessment of the applicants' high school academic performance (approximately 50%) and through a comprehensive review of the application including personal achievements of the applicant (approximately 50%) (University of California-Berkeley). For Fall 1999, the average weighted GPA of an admitted freshman was 4.16, with a SAT I verbal score range of 580-710 and a SAT I math score range of 620-730 for the middle 50% of admitted students (University of California-Berkeley). While there is no specific GPA for UC Berkeley applicants that will guarantee admission, a GPA of 2.8 or above is required for California residents and a test score total indicated in the University's Freshman Eligibility Index must be achieved. A minimum 3.4 GPA in A-F courses is required for non-residents. At Stanford University, most of the candidates have an un-weighted GPA between 3.6 and 4.0 and verbal SAT I and math SAT I scores of at least 650 (Stanford University Admission) At UC Berkeley, the academic assessment includes student's academic performance and several measured factors such as:

- College preparatory courses
- Advanced Placement (AP)
- International Baccalaureate Higher Level (IBHL)
- Honors and college courses beyond the UC minimum and degree of achievement in those courses
- Uncapped UC GPA
- Pattern of grades over time
- Scores on the three required SAT II tests and the SAT I (or ACT)
- Scores on AP or IBHL exams
- Honors and awards which reflect extraordinary, sustained intellectual or creative achievement
- Participation in rigorous academic enrichment
- Outreach programs
- Planned twelfth grade courses

- Qualification for UC Eligibility in the Local Context

All freshman applicants must complete courses in the University of California's A-F subject pattern and present scores from SAT I (or ACT) and SAT II tests with the following required subjects:

a. History/Social Science - 2 years required
b. English - 4 years required
c. Mathematics - 3 years required, 4 recommended
d. Laboratory Science - 2 years required, 3 recommended
e. Language Other than English - 2 years required, 3 recommended
f. College Preparatory Electives - 2 years required

At Stanford University, in addition to the academic transcript, close attention is paid to other factors such as student's written application, teacher references, the short responses and one-page essay (carefully read for quality, content, and creativity), and personal qualities.

The information provided in this study is a hypothetical situation and does not reflect the current UC system or Stanford University admissions criteria. However, we use this information to build a model to represent a real admissions problem. For more detailed information regarding University admissions, please refer to the University of California-Berkeley and Stanford University, Office of Undergraduate Admission (University of California-Berkeley; Stanford University Admission)

Given the factors above and the information contained in **Table 1**, a simulated-hypothetical model (a Virtual Model) was developed. A series of excellent, very good, good, not good, not bad, bad, and very bad student given the criteria for admission has been recognized. These criteria over time can be modified based on the success rate of students admitted to the university and their performances during the first, second, third and fourth years of their education with different weights and degrees of importance given for each year. Then, fuzzy similarity and ranking can evaluate a new student rating and find it's similarity to a given set of criteria.

Figure 2 shows a snapshot of the software developed for university admissions and the evaluation of student applications. **Table 1** shows the granulation of the

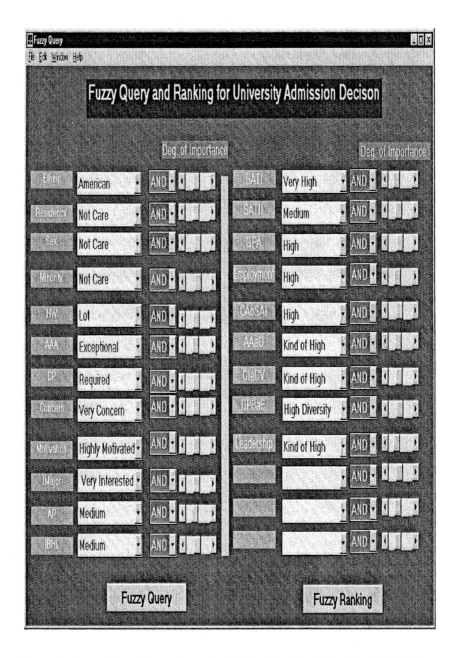

Figure 2. A snapshot of the software for University Admission Decision Making.

variables that was used in the model. To test the performance of the model, a demo version of the software is available at: http://zadeh.cs.berkeley.edu/ (Nikravesh, 2001a). Incorporating an electronic intelligent knowledge-based search engine, the results will eventually be in a format to permit a user to interact dynamically with the contained database and to customize and add information to the database. For instance, it will be possible to test an intuitive concept by dynamic interaction between software and the human mind.

This will provide the ability to answer "What if?" questions in order to decrease uncertainty and provide a better risk analysis to improve the chance for "increased success" on student selection or it can be used to select students on the basis of "diversity" criteria. The model can be used as for decision support and for a more uniform, consistent and less subjective and biased way. Finally, the model could learn and provide the mean to include the feedback into the system through time and will be adapted to the new situation for defining better criteria for student selection.

In this study, it has been found that ranking and scoring is a very subjective problem and depends on user perception (**Figure 2 and Figure 3**) and preferences in addition to the techniques used for the aggregation process which will effect the process of the data mining in reduced domain (**Figure 4**). Therefore, user feedback and an interactive model are recommended tools to fine-tune the preferences based on user constraints. This will allow the representation of a multi-objective optimization with a large number of constraints for complex problems such as credit scoring or admissions. To solve such subjective and multi-criteria optimization problems, GA-fuzzy logic and DNA-fuzzy logic models are good candidates.

In the case of the GA-Fuzzy logic model, the fitness function will be defined based on user constraints. For example, in the admissions problem, assume that we would like to select students not only on the basis of their achievements and criteria defined in Table 3, but also on the basis of diversity which includes gender distribution, ethnic background distribution, geophysical location distribution, etc. The question will be "what are the values for the preferences and which criteria should be used to achieve such a goal?" In this case, we will define the genes as the values for the preferences and the fitness function will be defined as the degree by which the distribution of each candidate in each generation match the desired distribution. fuzzy similarity can be used to define the degree of match which can be used for better decision analysis.

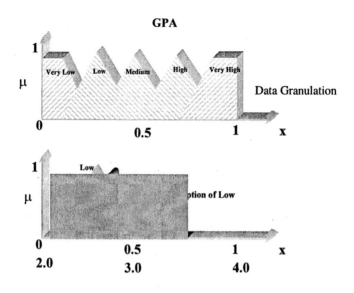

Figure 2. User's perception of "GPA Low"

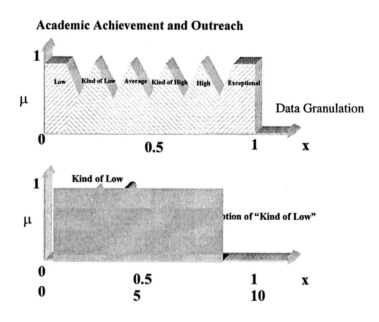

Figure 3. User's perception of Academic Achievement "Kid of Low"

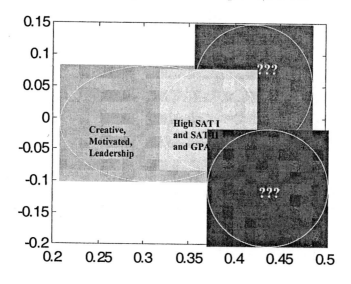

Figure 4. Typical Text and Rule Data Mining.

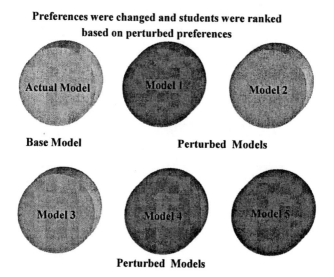

Figure 5. Models 1 through 5 are models based on preferences were perturbed around the actual value.

3.1 Effect of Preferences on Ranking of Students

To study the effect of preferences in the process of student selection and in the process of the ranking, the preferences in **Figure 1** were changed and students were ranked based on perturbed preferences, models 1 through 5 in **Figure 5.**

Figures 6.a through **6.d** show the results of the ranking of the students given the models 1 through 5. It is shown that given less than %10 changes on the actual preferences, most of the students were mis-ranked and mis-placed. Out of 100 students, less than %50 students or as an average only %41 of the actual students were selected (**Figure 6.a**). **Figure 6.b** shows that only less than %70 of the students will be correctly selected if we increase the admission by a factor of two, around %85 if we increase the admission by a factor of 3 (**Figure 6.c**), and less than %90 if we increase the admission by a factor of 4 (**Figure 6.d**). **Figures 7.a** through **7.d** show typical distribution of the 21 variables used for the Admission model. **Figures 7.a** through **7.d** show that the distribution of the students also drastically has been changed.

Now, the question will be "what are the values for the preferences and which criteria should be used to achieve such a goal?"

- Given a set of successful students, we would like to adjust the preferences such that the model could reflect this set of students.

- Diversity which includes gender distribution, ethnic background distribution, geophysical location distribution, etc.

To solve such subjective and multi-criteria optimization problems with a large number of constraints for complex problems such as University Admissions, the BISC Decision Support System is an excellent candidate.

4 BISC Decision Support System

Decision Support systems may represented in either of the following forms 1) physical replica of a system, 2) analog or physical model, 3) mathematical (qualitative) model, and 4) mental models. Decision support system is an approach or a philosophy rather than a precise methodology that can be used mainly for

- strategic planning such as resource allocation
- management control such as efficient resources utilization

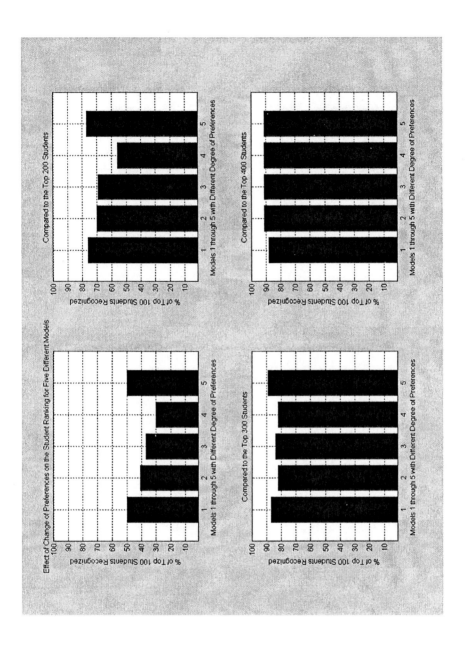

Figure 6. Effect of less than +-%10 Random perturbation on Preferences on the recognition of the pre-selected students given actual model.

201

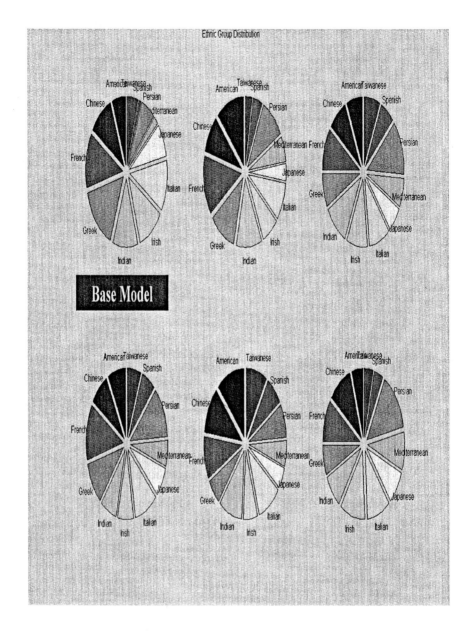

Figure 7.a. Ethnic Group Distribution

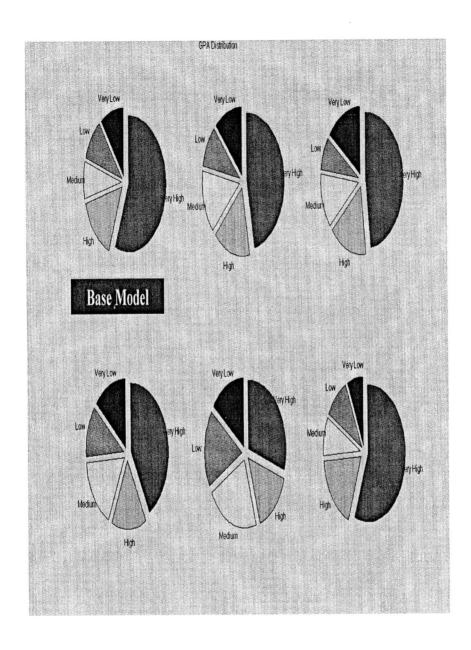

Figure 7.b. GPA Distribution

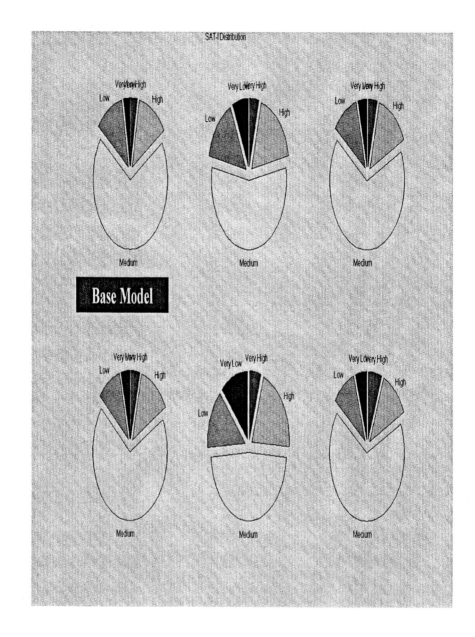

Figure 7.c. SAT-I Distribution

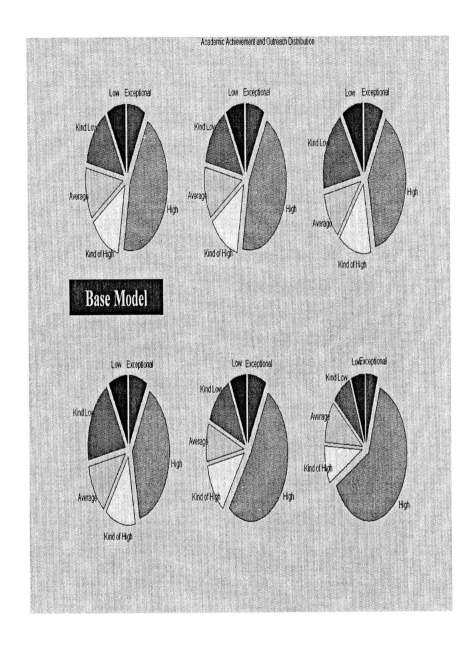

Figure 7.d. Academic Achievement Distribution

- operational control for efficient and effective execution of specific tasks

Decision support system is an approach or a strategy rather than a precise methodology, which can be used for 1) use intelligently the vast amounts of important data in organizations in an optimum way as a decision support system and 2) share intelligently and securely company's data internally and with business partners and customers that can be process quickly by end users and more specifically for :
- strategic planning such as resource allocation
- management control such as efficient resources utilization
- operational control for efficient and effective execution of specific tasks

The main key features of the Decision Support System for the internet applications are 1) to use intelligently the vast amounts of important data in organizations in an optimum way as a decision support system and 2) To share intelligently and securely company's data internally and with business partners and customers that can be process quickly by end users.

The BISC (Berkeley Initiative in Soft Computing) Decision Support System Components include (**Figure 8**):

- Data Management: database(s) which contains relevant data for the decision process

- User Interface
    o users and decision support systems (DSS) communication

- Model Management and Data Mining
    o includes software with quantitative and fuzzy models including aggregation process, query, ranking, and fitness evaluation

- Knowledge Management and Expert System: model representation including
    o linguistic formulation,
    o functional requirements
    o constraints
    o goal and objectives
    o linguistic variables requirements

- Evolutionary Kernel and Learning Process
    o Includes software with quantitative and fuzzy models including, Fuzzy-GA, fuzzy aggregation process, ranking, and fitness evaluation

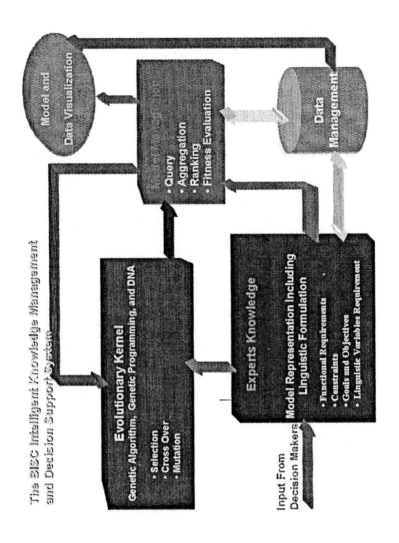

Figure 8. BISC Decision Support System

- Data Visualization: Allows end-users or decision makers can intervene in the decision-making process and see the results of the intervention

4.1 Implementation- BISC Decision Support System

In this section, we will introduce the BISC-DSS system for university admissions. In the case study, we used the GA-Fuzzy logic model for optimization purposes. The fitness function will be defined based on user constraints. For example, in the admissions problem, assume that we would like to select students not only on the basis of their achievements and criteria defined in Table 3 as a successful student, but also on the basis of diversity which includes gender distribution, ethnic background distribution, geophysical location distribution, etc. The question will be "what are the values for the preferences and which criteria should be used to achieve such a goal?" In this case, we will define the genes as the values for the preferences and the fitness function will be defined as the degree by which the distribution of each candidate in each generation match the desired distribution. Fuzzy similarity can be used to define the degree of match, which can be used for better decision analysis.

Figure 9 shows the performance of the conventional GA. The program has been run for 5000 generations and **Figure 9** shows the last 500 GA generations. As it is shown, the GA technique has been approached to a fitness of 80% and no further improvement was expected. Given what has been learned in each generation with respect to trends in the good genes, a series of genes were selected in each generation and has been used to introduce a new initial population to be used for GA. This process has been repeated until it was expected no improvement be achieved. **Figure 10** shows the performance of this interaction. The new model has reached a new fitness value, which is over 95%. **Figure 11** show the results of the ranking of the students given the actual model, predicted model (Model number 1) and models 2 through 4 which has been used to generate the initial population for training the fuzzy-GA model. It is shown that the predicted model ranked and selected most of the predefined students (**Figures 11.a-11.d**) and predefined distributions (**Figures 12.a-12.f**) and properly represented the actual model even

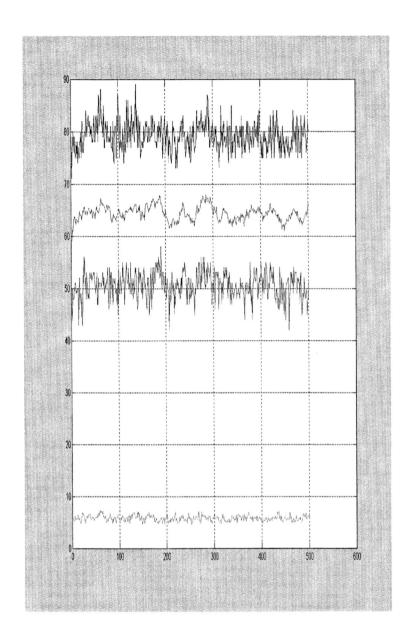

Figure 9. Conventional GA: Multi-Objective Multi-Criteria Optimization for
the University Admission

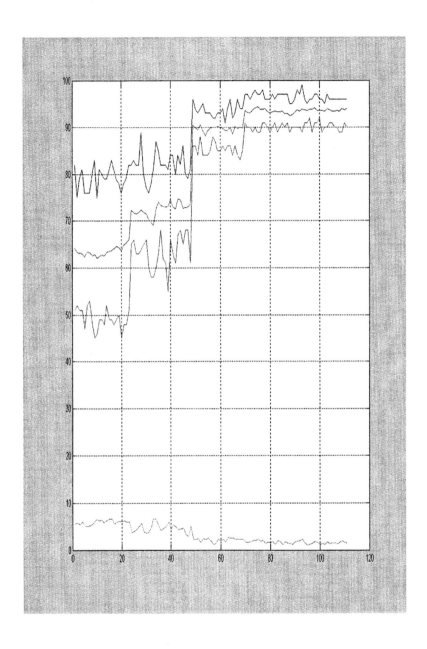

Figure 10. Interactive-GA Multi-Objective Multi-Criteria Optimization for the University Admission

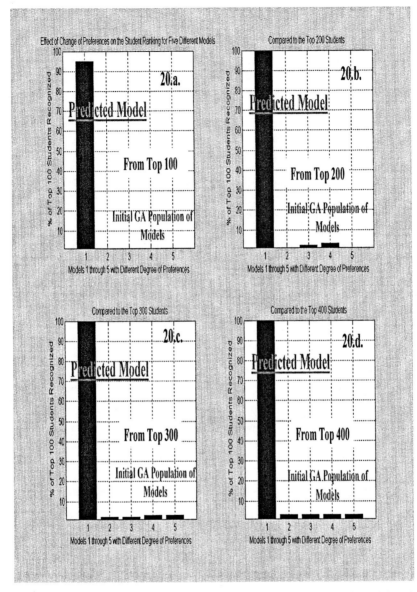

Figure 11. Results of the Ranking of the Students given Predicted Model and initial population for Fuzzy-GA Model

though the initial models to generate the initial population for training were far from the actual solution (**Figures 11.a-20.d and 12.a-12.f**). Out of 100 students, more than 90% students of the actual students were selected (**Figure 11.a**). **Figure 11.b** shows that %100 of the students will be correctly selected if we increase the admission by a factor of less than two. In has been concluded for this case study that %100 of students were selected if we increase the student admission by a factor of less than 1.15. **Figures 11.a-11.d** and **12.a-12.f** show that the initial models, model 2 through 5, were far from the actual model. Out of 100 students, less than 3% of the actual students were selected (**Figure 11.a**), around 5% if we increase the admission by a factor of 2 (**Figure 11.b**), around 10% if we increase the admission by a factor of 3 (**Figure 11.c**), and less than 15% if we increase the admission by a factor of 4 (**Fiure. 11.d**). **Figures 12.a-12.f** show typical distribution of the 21 variables used for the admission model. **Figures 12.a** through **12.f** show that the distribution of the student are properly presented by the predicted model and there is an excellent match between the actual model and the predicted model, even though the distributions of the initial populations are far from the actual model.

Figure 13 shows the results from data mining in reduced domain using part of a selected dataset as shown on **Figure 9** as a typical representation and techniques and strategy represented in **Figure 14**.

5 Other Potential Applications

In this section, we introduce fuzzy query and fuzzy aggregation for credit scoring, credit card ranking, and date matching applications.

5.1 Application to Credit Scoring (Fair, Isaac and Co)

Credit scoring was first developed in the 1950's and has been used extensively in the last two decades. In the early 1980's, the three major credit bureaus, Equitax, Experian, and TransUnion worked with the Fair Isaac Company to develop generic scoring models that allow each bureau to offer an individual score based on the contents of the credit bureau's data. FICO is used to make billions of financing decisions each year serving a 100 billion dollar industry. Credit scoring is a statistical method to assess an individual's credit worthiness and the likelihood that the individual will repay his/her loans based on their credit history and current credit accounts. The credit report is a snapshot of the credit history and the credit score is a snapshot of the risk at a particular point in time. Since 1995, this scoring system has made its biggest contribution in the world of mortgage lending. Mortgage investors such as Freddie Mac and Fannie Mae, the two main government-chartered

Typical Distribution of the Variables used for the Admission Model; Actual, Predicted and Initial Models for Fuzzy-GA

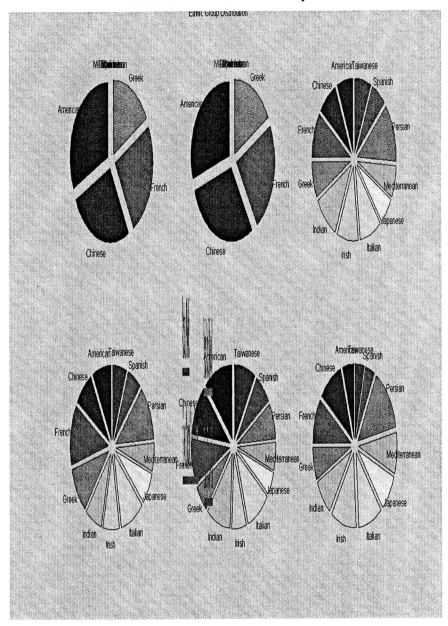

Figure 12.a. Ethnic Group Distribution

Typical Distribution of the Variables used for the Admission Model; Actual, Predicted and Initial Models for Fuzzy-GA

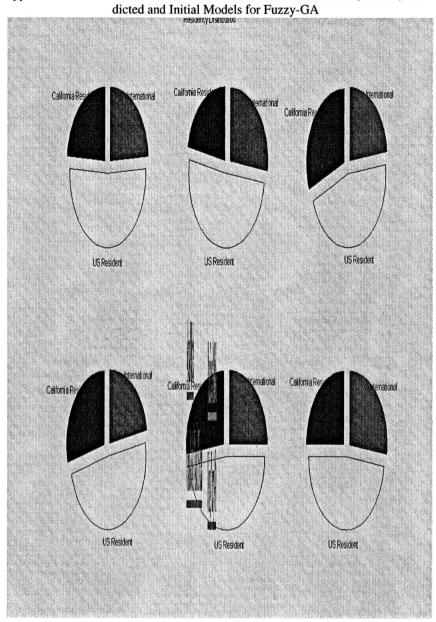

Figure 12.b. Residency Distribution

Typical Distribution of the Variables used for the Admission Model; Actual, Predicted and Initial Models for Fuzzy-GA

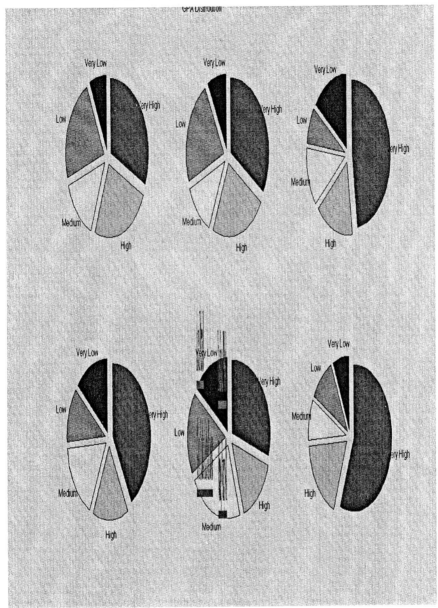

Figure 12.c. GPA Distribution

Typical Distribution of the Variables used for the Admission Model; Actual, Predicted and Initial Models for Fuzzy-GA

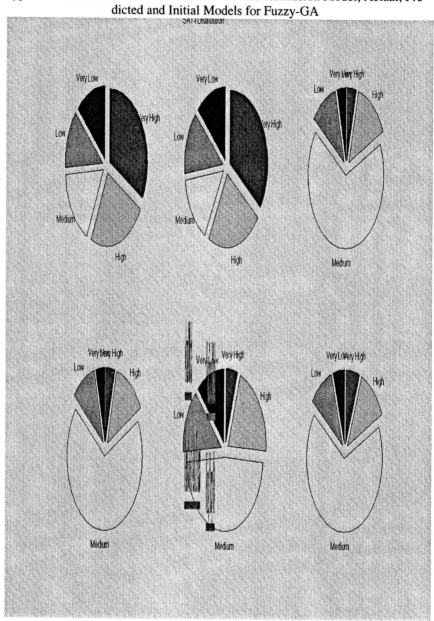

Figure 12.d. SAT-I Distribution

Typical Distribution of the Variables used for the Admission Model; Actual, Predicted and Initial Models for Fuzzy-GA

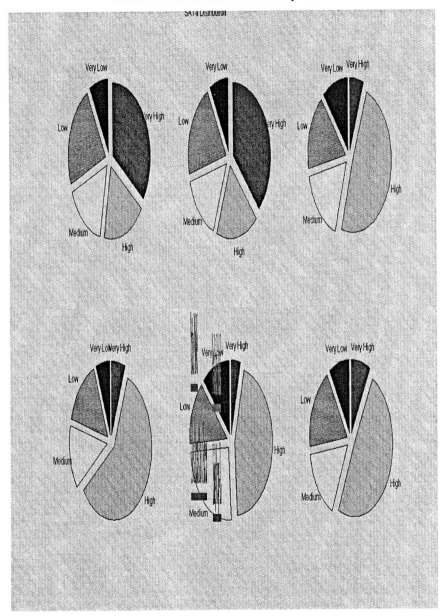

Figure 12.e. SAT-II Distribution

Typical Distribution of the Variables used for the Admission Model; Actual, Predicted and Initial Models for Fuzzy-GA

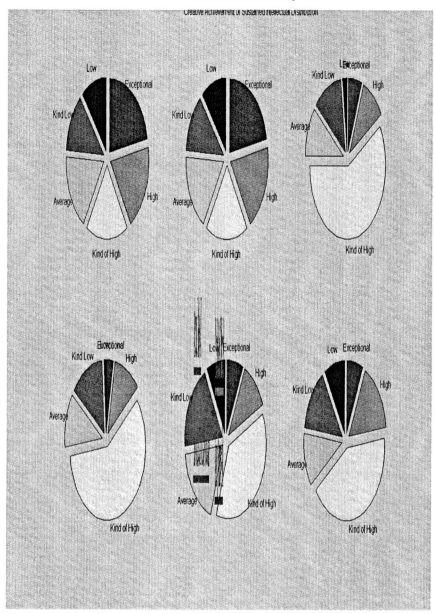

Figure 12.f. Creative Achievement or Sustained Intellectual Distribution

Data Mining in Reduced Domain

Each Point Represents One Student

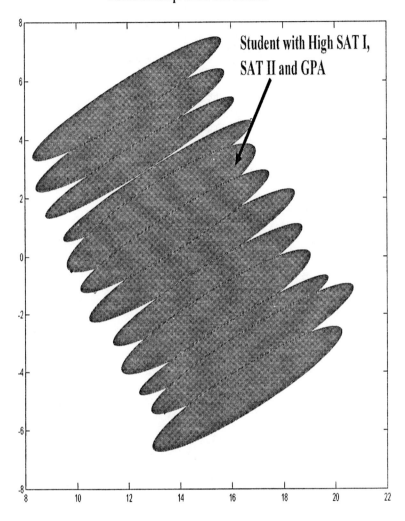

Student with High SAT I, SAT II and GPA

Figure 13. Data Mining based on Techniques described in "Search Strategy" and Figure 14. on selected dataset

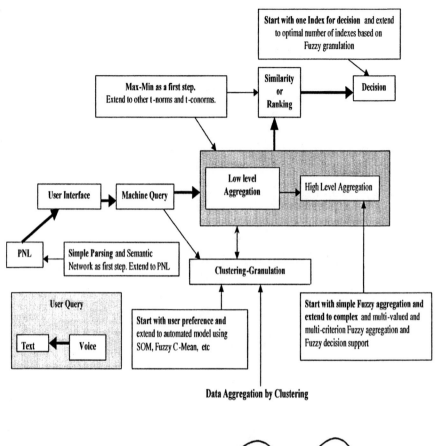

Figure 14. Simplified flow diagram and flow of information for PNL-Based Fuzzy Query.

companies that purchase billion of dollars of newly originated home loans annually, endorsed the Fair Isaac credit bureau risk, ignored subjective considerations, but agreed that lenders should also focus on other outside factors when making a decision.

When you apply for financing, whether it's a new credit card, car or student loan, or a mortgage, about 40 pieces of information from your credit card report are fed into a model. This information is categorized into the following five categories with different level of importance (% of the score):

1. Past payment history (35%)

2. Amount of credit owed (30%)

3. Length of time credit established (15%)

4. Search for and acquisition of new credit (10%)

5. Types of credit established (10%)

When a lender receives your Fair Isaac credit bureau risk score, up to four "score reason codes" are also delivered. These explain the reasons why your score was not higher. Followings are the most common given score reasons;

1. Serious delinquency

2. Serious delinquency, and public record or collection filed

3. Derogatory public record or collection filed

4. Time since delinquency is too recent or unknown

5. Level of delinquency on accounts

6. Number of accounts with delinquency

7. Amount owed on accounts

8. Proportion of balances to credit limits on revolving accounts is too high

9. Length of time accounts have been established

10. Too many accounts with balances

By analyzing a large sample of credit file information on people who recently obtained new credit, and given the above information, a statistical model has been built. The model provides a numerical score designed to predict your risk as a borrower. Credit scores used for mortgage lending range from 0 to 900 (usually above 300). The higher your score, the less risk you represent to lenders. Most lenders will be happy if your score is 700 or higher. You may still qualify for a loan with a lower score given all other factors, but it will cost you more. For example, given a score of around 620 and a $25,000 car loan for 60 months, you will pay approximately $4,500 more than with a score of 700. You will pay approximately $6,500 more than if your score is 720. Thus, a $25,000 car loan for 60 months with bad credit will cost you over $10,000 more for the life of the loan than if you have an excellent credit score.

5.2 Application to Credit Card Ranking (U.S. Citizens for Fair Credit Card Terms)

Credit ratings that are compiled by the consumer credit organization such as the U.S. Citizens for Fair Credit Card Terms (CFCCT) could simply save you hundreds of dollars in credit card interest or help you receive valuable credit card rebates and rewards including frequent flyer miles (free airline tickets), free gas, and even hundreds of dollars in cash back bonuses.

CFCCT has developed an objective-based method for ranking credit cards in US. In this model, interest rate has the highest weighting in the ranking formula. FCC rates credit cards based on the following criteria:

1. Purchase APR
2. Cash Advance APR
3. Annual Fees
4. Penalty for cards that begin their grace periods at the time of purchase/posting instead of at the time of billing
5. Bonuses for cards that don't have cash advance fees
6. Bonuses for cards that limit their total cash advance fees to $10.00
7. Bonuses for introductory interest rate offers for purchases and/or balance transfers
8. Bonuses for cards that have rebate/perk programs
9. Bonuses for cards that have fixed interest rates.

We used the top 10 classic cards, the top 10 gold cards, and the top 10 platinum cards which have been ranked by the CFCCT method as of March 2001. Given the above factors, a simulated model has been developed. A series of excellent, very good, good, not good, not bad, bad, and very bad credit cards have been recognized for the credit cards selected. Then, fuzzy similarity and ranking has been used to rank the cards and define a credit score.

5.3 Date Matching

The main objective of this project was to find the best possible match in the huge space of possible outputs in the databases using the imprecise matching such as fuzzy logic concept, by storing the query attributes and continuously refining the query to update the user's preferences. We have also built a Fuzzy Query system, which is a java application that sits on top of a database.

With traditional SQL queries (relational DBMS), one can select records that match the selection criteria from a database. However, a record will not be selected if any one of the conditions fails. This makes searching for a range of potential candidates difficult. For example, if a company wants to find an employee who is proficient in skill A, B, C and D, they may not get any matching records, only because some candidates are proficient in 3 out of 4 skills and only semi-proficient in the other one. Since traditional SQL queries only perform Boolean matching, some qualities of real life, like "far" or "expensive" or "proficient", which involve matters of degree, are difficult to search for in relational databases. Unlike Boolean logic, fuzzy logic allows the degree of membership for each element to range over an interval. So in a fuzzy query, we can compute how similar a record in the database is to the desired record. This degree of similarity can be used as a ranking for each record in the database. Thus, the aim of the fuzzy query project for date matching is to add the capability of imprecise querying (retrieving similar records) to traditional DBMS. This makes some complex SQL statements unnecessary and also eliminates some repetitious SQL queries (due to empty-matching result sets).

In this program, one can basically retrieve all the records from the database, compare them with the desired record, aggregate the data, compute the ranking, and then output the records in the order of their rankings. Retrieving all the records from the database is a naïve approach because with some preprocessing, some very different records are not needed from the database. However, the main task is to compute the fuzzy rankings of the records so efficiency is not the main concern here.

The major difference between this application and other date matching system is that a user can input his hobbies in a fuzzy sense using a slider instead of choosing crisp terms like "Kind of" or "Love it". These values are stored in the database according to the slider value (**Figures 15 and 16**) .

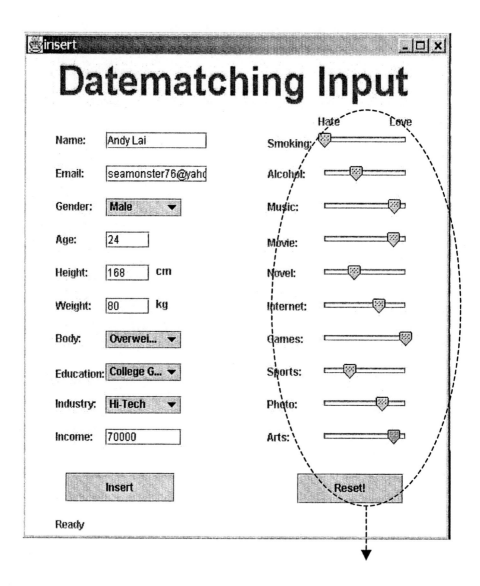

Figure 16. Date matching nput form

224

Desired Fuzzy Attributes, which are similar to those in the data, input menu. However, these can be replaced by selection menu here.

Desired
Attributes

A user can input how importance an attribute is to the Fuzzy Query. Degree 0 means don't care.

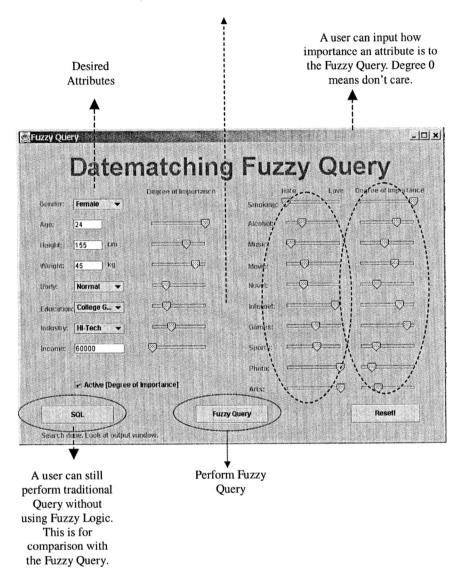

A user can still perform traditional Query without using Fuzzy Logic. This is for comparison with the Fuzzy Query.

Perform Fuzzy Query

Figure 16. Snapshot of the Date Matching Software

Figure 17 shows the results are obtained from fuzzy query using the search criteria in the previous page. The first record is the one with the highest ranking – 80%. Note that it matches the age field of the search criteria but it's off a bit from the height and weight fields. So one can do imprecise querying.

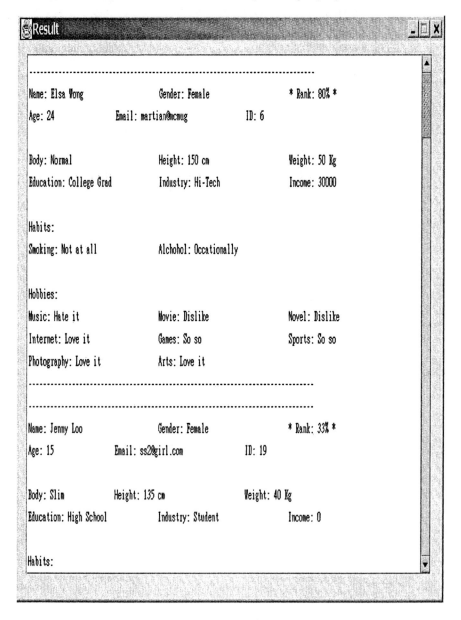

Figure 17. Sample of the output from Date Matching software

The system is modulated into three main modules **(Figure 18)**. The core module is the fuzzy engine which accepts input from a GUI module and outputs result to another GUI module. The GUIs can be replaced by other processing modules such that the input can be obtained from other system and the result can be used for further analysis.

High level structure of the project

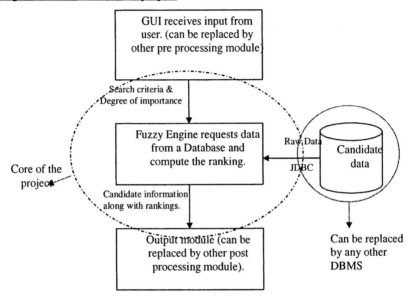

Figure 18. System Structure

The current date matching software can be modified or expanded in several ways:

1. One can build a server/client version of date-matching engine so that we can use a centralized database and all users around the world can do the matching through the web. The ranking part (computation) can still be done on local machine since every search is different. This can also help reduce the server load.

2. The attributes, granulation models and the "meaning" of the data can be tunable so that the system is more configurable and adaptive to changes.

227

3. User preference capability can be added to the system. (The notion of "overweight" and "tall" can be different to different people.)

4. The GUI needs to be changed to meet real user needs.

5. One can build a library of fuzzy operators and aggregation functions such that one can choose the operator and function that matches the application.

6. One can instead build a generic fuzzy engine framework which is tunable in every way to match clients' needs.

7. The attributes used in the system are not very complete compared to other data matching systems online. However, the attributes can be added or modified with some modification to the program without too much trouble.

Recently, we have added a web interface to the existing software and built the database framework for further analysis in user profiling so that users could find the best match in the huge space of possible outputs. We saved user profiles and used them as basic queries for that particular user. Then, we stored the queries of each user in order to "learn" about this user's preference. In addition, we rewrote the fuzzy search engine to be more generic so that it would fit any system with minimal changes. Administrator can also change the membership function to be used to do searches. Currently, we are working on a new generic software to be developed for a much more diverse applications and to be delivered as stand alone software to both academia and businesses.

6 Conclusions

In this study, we introduced fuzzy query and fuzzy aggregation and the BISC decision support system as an alternative for ranking and predicting the risk for credit scoring, university admissions, and several other applications which currently utilize an imprecise and subjective process. The BISC decision support system key features are 1)intelligent tools to assist decision-makers in assessing the consequences of decision made in an environment of imprecision, uncertainty, and partial truth and providing a systematic risk analysis, 2)intelligent tools to be used to assist decision-makers answer "What if Questions", examine numerous alternatives very quickly and find the value of the inputs to achieve a desired level of output, and 3) intelligent tools to be used with human interaction and feedback to achieve a capability to learn and adapt through time In addition, the following important points have been found in this study 1) no single ranking function works well for all contexts, 2) most similarity measures work about the same regardless of the model, 3) there is little overlap between successful ranking functions, and 4) the same model can be used for other applications such as the design of a more intelligent search engine which includes the user's preferences and profile (Nikravesh 2001a and 2001b).

Acknowledgement

Funding for this research was provided by the British Telecommunication (BT) and the BISC Program of UC Berkeley. Authors would like to thank the former EECS-UC Berkeley student, Hoi Kit Lai (CS 199 project), for development of the Date Matching software and providing the information given in section 5.3.

References

1. Fair, Isaac and Co.: http://www.fairisaac.com/.
2. Bonissone P.P., Decker K.S. (1986) Selecting Uncertainty Calculi and Granularity: An Experiment in Trading; Precision and Complexity, in Uncertainty in Artificial Intelligence (L. N. Kanal and J. F. Lemmer, Eds.), Amsterdam.
3. Fagin R. (1998) Fuzzy Queries in Multimedia Database Systems, Proc. ACM Symposium on Principles of Database Systems, pp. 1-10.
4. Fagin R. (1999) Combining fuzzy information from multiple systems. J. Computer and System Sciences 58, pp 83-99.
5. Mizumoto M. (1989) Pictorial Representations of Fuzzy Connectives, Part I: Cases of T-norms, T-conorms and Averaging Operators, Fuzzy Sets and Systems 31, pp. 217-242.
6. Nikravesh M. (2001a) Perception-based information processing and retrieval: application to user profiling, 2001 research summary, EECS, ERL, University of California, Berkeley, BT-BISC Project. (http://zadeh.cs.berkeley.edu/ & http://www.cs.berkeley.edu/~nikraves/ & http://www-bisc.cs.berkeley.edu/).
7. Nikravesh M. (2001b) Credit Scoring for Billions of Financing Decisions, Joint 9th IFSA World Congress and 20th NAFIPS International Conference. IFSA/NAFIPS 2001 "Fuzziness and Soft Computing in the New Millenium", Vancouver, Canada, July 25-28, 2001.
8. Stanford University Admission, http://www.stanford.edu/home/stanford/facts/undergraduate.html
9. U.S. Citizens for Fair Credit Card Terms; http://www.cardratings.org/cardrepfr.html.
10. University of California-Berkeley, Office of Undergraduate Admission, http://advising.berkeley.edu/ouars/.

Web Mining in Soft Computing Framework: A Survey

Sankar K. Pal[1], Varun Talwar[2], and Pabitra Mitra[1]

[1] Machine Intelligence Unit
Indian Statistical Institute
Kolkata 700108, India
E-mail: {sankar,pabitra_r}@isical.ac.in
[2] Department of Computer Science
National University of Singapore
Singapore 119260
E-mail: varuntal@comp.nus.edu.sg

Summary. The chapter deals with use of different soft computing tools to achieve web intelligence. It summarizes different characteristics of web data, the basic components of web mining and its different types, and their current states of the art. The reason for considering web mining, a separate field from data mining, is explained. The limitations of some of the existing web mining methods and tools are enunciated, and the significance of soft computing (comprising fuzzy logic, artificial neural networks, genetic algorithms and rough sets) highlighted. A survey of the existing literature on 'soft web mining' is provided along with the commercially available systems. The prospective areas of web mining where the application of soft computing needs immediate attention are outlined with justification. Scope for future research in developing 'soft web mining' systems is explained. An extensive bibliography is also provided.

Keywords: Data mining, knowledge discovery, information retrieval, search engines, pattern recognition, fuzzy logic, artificial neural networks, rough sets, genetic algorithms

1 Introduction

The evolution of the internet into the global information network, coupled with the immense popularity of the web as a means of information dissemination, has lead to an explosive growth in the information available on the World Wide Web (WWW). Given that there is this vast and ever growing amount of information, how does the average user quickly find what s/he is looking for – a task in which the present day search engines don't seem to help much! This has prompted the need for developing automatic mining techniques on WWW, thereby giving rise to the term 'web mining'.

To proceed towards web intelligence, obviating the need for human intervention, we need to incorporate and embed artificial intelligence into web tools. The necessity of creating server-side and client-side intelligent systems that can effectively mine for knowledge both across the internet and in particular web localities is drawing the attention of researchers from the domains of information retrieval, knowledge discovery, machine learning, and AI, among others. However, the problem of developing automated tools in order to find, extract, filter, and evaluate the users desired information from unlabelled, distributed and heterogeneous web data is far from being solved. To handle these characteristics and overcome some of the limitations of existing methodologies, soft computing seems to be a good candidate; the research area combining the two may be termed as 'soft web mining' .

At present the principal soft computing tools include, fuzzy sets, artificial neural networks, genetic algorithms and rough set theory. Fuzzy sets provide a natural framework for the process in dealing with uncertainty. Neural networks are widely used for modelling complex functions, and provide learning and generalization capabilities. Genetic algorithms (GAs) are an efficient search and optimization tool. Rough sets help in granular computation and knowledge discovery.

The objective of this chapter is to provide an outline of web mining, its various classifications, its subtasks and to give a perspective to the research community about the potential of applying soft computing techniques to its different components. The article, besides reviewing the existing techniques/tools and their limitations, lays emphasis on possible enhancements of these tools using soft computing framework. In this regard, the relevance of fuzzy logic, artificial neural networks, genetic algorithms and rough sets is illustrated through examples and diagrams. Broad guidelines for future research, on web mining, in general, are outlined.

2 Web Mining

The web is a vast collection of completely uncontrolled heterogeneous documents. Thus it is huge, diverse and dynamic, and raises the issues of scalability, heterogeneity and dynamism respectively. Due to these characteristics we are currently drowning in information, but starving for knowledge; thereby making the web a fertile area of data mining research with the huge amount of information available online. Data mining refers to the non trivial process of identifying valid, novel, potentially useful and ultimately understandable patterns in data.

Web mining can be broadly defined as the discovery and analysis of useful information from the World Wide Web (WWW). In web mining data can be collected at the server side, client side, proxy servers or obtained from an organization's database. Depending on the location of the source, the type of collected data differs. It also has extreme variation both in its content (e.g., text, image, audio, symbolic) and meta information, that might be available. This makes the techniques to be used for a particular task in web mining widely varying. Some of the characteristics of web data are:

(a) unlabelled
(b) distributed
(c) heterogeneous (mixed media)
(d) semi-structured
(e) time varying
(f) high dimensional

Therefore, web mining basically deals with mining large and hyper-linked information base having the aforesaid characteristics. Also being an interactive medium, human interface is a key component of most web applications. Some of the issues which have come to light, as a result, concern:

(a) need for handling context sensitive and imprecise queries,
(b) need for summarization and deduction,
(c) need for personalization and learning.

Thus web mining, though considered to be a particular application of data mining, warrants a separate field of research mainly because of the aforesaid characteristics of the data and human related issues.

2.1 Web Mining Components and the Methodologies

Web mining can be viewed as consisting of four tasks, shown in Fig. 1. Each task is described below along with a survey of the existing methodologies/ tools for the task.

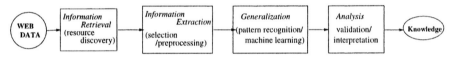

Fig. 1. Web mining subtasks

Information Retrieval (Resource Discovery):

Resource discovery or information retrieval (IR) deals with automatic retrieval of all relevant documents, while at the same time ensuring that the non-relevant ones are fetched as few as possible. The IR process mainly includes document representation, indexing and searching for documents.

Different techniques used by various authors for document representation in information retrieval of web content mining (to be explained in detail in Section 3) for semi structured documents are surveyed in [36]. An index is, basically, a collection of terms with pointers to places where the information about documents can be found. However, indexing of web pages to facilitate retrieval is quite a complex and challenging problem as compared to the corresponding one associated with classical databases where straightforward techniques suffice. The enormous number of pages on the web, their dynamism and frequent updation make the indexing techniques seemingly impossible. At present, four approaches to index documents on the web are: human or manual indexing, automatic indexing, intelligent or agent-based indexing and metadata based indexing.

Search engines are programs written to query and retrieve information stored in databases (fully structured), HTML pages (semi structured) and free text (unstructured) on the web. The most popular search indices have been created by web robots such as AltaVista and WebCrawler which scan millions of web documents and store an index of the words in the documents. There are over a dozen different indices currently in active use, each with a unique interface and a database covering a different fraction of the web. MetaCrawler presents the next level of information food chain by providing a single unified interface for web document searching [14]. It submits the query to nine indices in parallel and then collates the results and prunes them. Thus instead of tackling the web directly, MetaCrawler mines robot-created searchable indices. Future resource discovery systems will make use of automatic text categorization technology to classify web documents into categories. This technology could facilitate the automatic construction of web directories as well as filtering of the results of queries to searchable indices. Besides document indexing and categorization, ranking of a document in terms of its importance is an important aspect of search technology. Different strategies are used by individual engines. The most popular among them is perhaps the PageRank technique used by Google [5]. Fig. 2 shows the architecture of the Google search engine.

Information Selection/Extraction and Preprocessing:

Once the documents have been retrieved the challenge is to automatically extract knowledge and other required information without human interaction. Information Extraction (IE) is the task of identifying specific fragments of a single document that constitute its core semantic content. Till now the major methods of IE involve writing *wrappers* (hand coding) which map the documents to some data model. Another method for IE from hypertext is given in [18] where each page is approached with a set of standard questions. The problem, therefore, reduces to identifying the text fragments which answer those specific questions.

Scalability is the biggest challenge to IE experts; it is not feasible to build IE systems which are scalable to the size and dynamism of the web. Another challenge is integration and preprocessing of heterogeneous data. When a user requests a web page, a variety of files like images, sound, video, executable cgi and html, are accessed. As a result, the server log contains many entries that are redundant or irrelevant for mining tasks. Therefore these need to be removed through preprocessng. One of the preprocessing techniques used for IE is Latent Semantic Indexing (LSI) that seeks to transform the original document vectors to a lower dimensional space by analyzing the correlational structure of terms in that document collection such that similar documents that do not share the same terms are placed in the same category (topic). 'Stemming' is yet another preprocessing technique which reduces the input feature size by stemming words like *informed, information, informing* to their root *inform*.

It may be noted that IE aims to extract new knowledge from the retrieved documents by capitalizing on the document structure and representation of the document, whereas IR experts view the document text just as a bag of words and do not pay attention to the structure of the document.

Generalization:

In this phase, pattern recognition and machine learning techniques are usually used on the extracted information . Most of the machine learning systems, deployed on the web, learn more about the user's interest than the web itself. A major obstacle when learning about the web is the labelling problem: data is abundant on the web but it is unlabelled. Many data mining techniques require inputs labelled as positive (yes) or negative (no) examples with respect to some concept. For example, if we are given a large set of web pages labelled as positive and negative examples of the concept *homepage*, then it is easy to design a classifier that predicts whether any unknown web page is a home page or not; unfortunately web pages are unlabelled. Techniques such as uncertainty sampling reduce the amount of unlabelled data needed, but do not eliminate the labelling problem. An approach to solve this problem is based on the fact that the web is much more than just a linked collection of documents, it is an interactive medium. For example, 'Ahoy' [16] takes as input a person's name and affiliation, and attempts to locate the person's homepage; hence it asks the users to label its answers as correct or incorrect. Clustering techniques do not require labelled inputs and have been applied successfully to large collections of documents [12]. Indeed, the web offers a fertile ground for document clustering research. Association rule mining is also an integral part of this phase. Basically, association rules are expressions of the type $X \Rightarrow Y$ where X and Y are sets of items. $X \Rightarrow Y$ expresses that whenever a transaction T contains X then T probably contains Y also. The probability or rule confidence is defined as the percentage of transactions containing Y in addition to X as compared to overall transactions containing X. The idea of mining association rules originates from the market based data where rules like "A customer who buys product $x1$ and $x2$ will also buy product y with probability $c\%$" are found [27].

Machine processable documents have led to the development of the concept of the 'Semantic web' which is inspired from the fact that most information on the web is designed for human consumption, and even if it was derived from a database with well defined meanings (in at least some terms) for its columns, the structure of the data is not evident to a robot browsing the web. Leaving aside the artificial intelligence problem of training machines to behave like people, the semantic web approach, instead, develops languages for expressing information in a machine processable form.

Analysis:

Analysis is a data driven problem which presumes that there is sufficient data available so that potentially useful information can be extracted and analyzed. Humans play an important role in the information or knowledge discovery process on the web since the web is an interactive medium. This is especially important for validation and or interpretation of the mined patterns which take place in this phase. Once the patterns have been discovered, analysts need appropriate tools like, Webviz system [63], to understand, visualize and interpret these patterns. Some others use OLAP (Online analytical processing) techniques such as data cubes for the purpose of simplifying the analysis of usage statistics from sever access logs. The WEBMINER [50] system proposes a structured query language (SQL)- like querying mechanism for querying the discovered knowledge (in the form of association rules and sequential patterns). Cohen [9] focuses on the nature of knowledge that one can derive from the web.

Based on the aforesaid four phases (Fig. 1), *web mining can be viewed as the use of data mining techniques to automatically retrieve, extract and evaluate information for knowledge discovery from web documents and services. Here evaluation includes both 'generalization' and 'analysis'.*

2.2 Web Mining Categories

Web mining may be of three types, namely, Web Content Mining (WCM), Web Structure Mining (WSM) and Web Usage Mining (WUM). Let us now describe them.

Web Content Mining:

WCM deals with the discovery of useful information from the web contents/ data/ documents/ services. Various kinds of document representation (used for IR) and the tasks performed (for generalization) in WCM are described in [36]. However, web contents are not only text, but encompass a very broad range of data such as audio, video, symbolic, metadata and hyperlinked data. Out of these, research at present is mostly centered around text and hypertext contents. The web text data can be of three types: (a) unstructured data such as free text (b) semi structured data such as HTML (c) fully structured data such as in tables or databases. Mining (a) is termed as KDT (knowledge discovery in texts) or text data mining or text mining. Text mining is a well developed subject and full coverage of it is beyond the scope of this survey. For details one may refer to [48]. For mining from symbolic knowledge extracted from the web, one may refer to [23]. Hypertext mining involves mining semi-structured HTML pages which have hyperlinks, besides text. In hypertext mining, at times, supervised learning or classification plays a key role, like in email, newsgroup management and maintaining web directories. For a good tutorial on hypertext mining one may refer to [6].

Web content mining can take two approaches: agent-based approach and database approach. The agent-based approach to web mining involves the development of sophisticated AI systems that can act autonomously or semi-autonomously on behalf of a particular user, to discover and organize web-based information. Generally, agent-based web mining approach can be further organized into three categories: intelligent search agents, information filtering/ categorization and personalized web agents [10]. The database approach focuses on techniques for organizing semi-structured data on the web into more structured collection of resources, and uses database querying mechanisms and data mining techniques to analyze it. Levy et. al., [40] discuss general intelligent internet systems with respect to user modelling, discovery and analysis of remote information sources, information integration and web-site management.

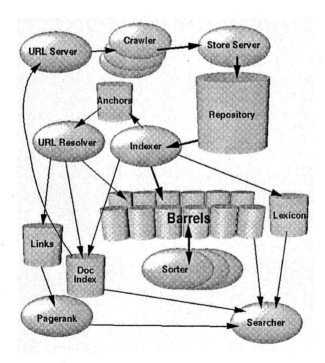

Fig. 2. Google © architecture

237

Web Structure Mining:

Web structure mining pertains to mining the structure of hyperlinks within the web itself (inter document structure unlike web content mining which pertains to intra document structure). Here, structure represents the graph of the links in a site or between sites. For example, Fig. 3 shows a sample link structure and levels of the site 'wharton.upenn.edu'. WSM reveals more information than just the information contained in documents. For example, links pointing to a document indicate the popularity of the document, while links coming out of a document indicate the richness or perhaps the variety of topics covered in the document. This was first highlighted and used in the HITS [33] algorithm. This is analogous to bibliographical citations. When a paper is cited often, it ought to be important. The link topology of the web has also been exploited to develop a notion of hyper linked communities. The analysis shows that communities can be viewed as containing a core of central "authoritative" pages linked together by hub pages; and they exhibit a natural type of hierarchical topic generalization that can be inferred directly from the pattern linkage. It shows that the notion of community provides a surprisingly clear perspective from which to view the seemingly haphazard development of web infrastructure [24]. The PageRank [5] and HITS [33] methods take advantage of this information conveyed by the links to find pertinent web pages. Focused Crawling [7] is a further enhancement in the field of hypertext resource discovery system. The goal of a focused crawler is to selectively seek out pages that are relevant to a pre-defined set of topics. The topics are specified not using keywords, but using exemplary documents. Rather than collecting and indexing all accessible web documents to be able to answer all possible ad-hoc queries, a focused crawler analyzes its crawl boundary to find the links that are likely to be most relevant for the crawl, and avoids irrelevant regions of the Web. This leads to significant savings in hardware and network resources, and helps to keep the crawl more up-to-date.

Web Usage Mining:

While content mining and structure mining utilize the real or primary data on the web, usage mining mines secondary data generated by the users' interaction with the web. Web usage data includes data from web server access logs, proxy server logs, browser logs, user profiles, registration files, user sessions or transactions, user queries, bookmark folders, mouse-clicks and scrolls, and any other data generated by the interaction of users and the web. WUM works on user profiles, user access patterns and mining navigation paths which are being heavily used by e-commerce companies for tracking customer behavior on their sites. WUM plays a key role in personalizing space, which is the need of the hour. To satisfy all the users with the same tool is extremely difficult, and, instead, we need to learn user access patterns, their path patterns at an individual level and also as a whole at web sites. Besides learning access

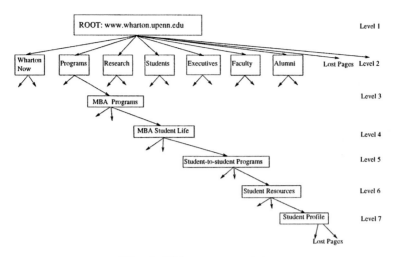

Fig. 3. Web structure mining

patterns, one needs to use 'collaborative filtering' for listing other users with similar interests. Collaborative recommender systems allow personalization for e-commerce by exploiting similarities and dissimilarities in users' preferences. A new algorithm is suggested in [42] for specifically catering to association rule mining in collaborative recommendation systems. In [60] a framework is given for applying machine learning algorithms along with feature reduction techniques, such as Singular Value Decomposition (SVD), for collaborative recommendation. It uses feature reduction techniques to reduce the dimension of the rating data and then neural networks are applied on the simplified data to make a model for collaborative recommendation. However, the discovery of patterns from usage data by itself is not sufficient for performing personalization tasks. A way of deriving good quality and useful 'aggregate user profiles' from patterns is suggested in [49]. It evaluates two techniques based on clustering of user transactions and clustering of page views, in order to discover overlapping aggregate profiles that can be effectively used by recommender systems for real time personalization. A framework for web mining has been proposed in WEBMINER [50] for pattern discovery from WWW transactions. Fig. 4 shows the WEBMINER architecture. It clearly brings out that, after integrating registration data and 'cleaned' transactions data, pattern discovery techniques are applied to discover association rules, sequential patterns and clusters, and classification rules.

3 Limitations of Existing Web Mining Methods

The web creates new challenges to different component tasks of web mining (Fig. 1) as the amount of information on the web is increasing and changing

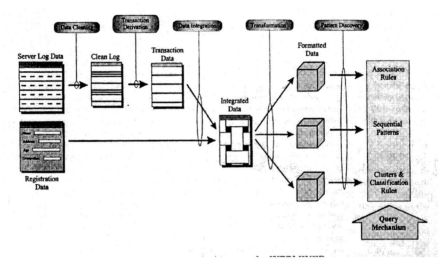

Fig. 4. Webminer architecture

rapidly without any control. As a result, the existing systems find difficulty in handling the newly emerged problems during information retrieval, information extraction, generalization (clustering and association) and analysis. Some of these are described below.

3.1 Information Retrieval

The following difficulties may be encountered during this task:

Subjectivity, Imprecision and Uncertainty: The aim of an IR system is to estimate the relevance of documents to users' information needs, expressed by means of queries. This is a hard and complex task which most of the existing IR systems find difficult to handle due to the inherent subjectivity, imprecision and uncertainty related to user queries. Most of the existing IR systems offer a very simple modelling of retrieval, which privileges the efficiency at the expense of accuracy. Query processing in search engines, which are an important part of IR systems, is simple blind keyword matching. This does not take into account the context and relevance of queries with respect to documents, while these are important for efficient machine learning.

Deduction: The current search engines have no deductive capability. For example, none of them gives a satisfactory response to a query like: "How many computer science graduates were produced by European universities in 1999?"

Soft Decision: Current query processing techniques follow the principle of hard rejection while determining the relevance of a retrieved document with respect to a query. This is not correct since relevance, itself, is a 'gradual' property of the documents [53], not a crisp one.

240

Page Ranking: Page ranks are important since human beings find it difficult to scan through the entire list of documents returned by the search engine in response to his/her query. Rather, one sifts through only the first few pages, say less than twenty, to get the desired documents. Therefore, it is desirable, for convenience, to get the pages ranked with respect to 'relevance' to user queries.

However there is no definite formula which truly reflects such relevance in top ranked documents. The scheme for determining page ranks should incorporate (a) weights given to various parameters of the hit like location,proximity and frequency, (b) weight given to reputation of a source i.e. a link from yahoo.com should carry a much higher weight than link from any other not so popular site, and (c) ranks relative to the user.

Personalization: It is necessary that IR systems tailor the retrieved document set as per users' history or nature. Though some of the existing systems do so for a few limited problem domains, no definite general methodology is available. Although efforts in this direction have been made by clustering logged data, the similarity metric used in clustering is not meaningful and the principle on which it is derived is not clear.

Dynamism, Scale and Heterogeneity: IR systems find difficulty in dealing with the problem of dynamism, scaling and heterogeneity of web documents. Because of time varying nature of web data many of the documents returned by the search engines are outdated, irrelevant and unavailable in future, and hence the user has to try his queries across different indices several times before getting a satisfactory response. Regarding the scaling problem, Etzioni [17] has studied the effect of data size on precision of the results obtained by the search engine. Current IR systems are not able to index all the documents present on the web and this leads to the problem of 'low recall'. The heterogeneity nature of web documents demands a separate mining method for each type of data.

3.2 Information Extraction

Most of the IE algorithms used by different tools are based on the 'wrapper' technique. Wrappers are procedures for extracting a particular information from web resources. Its biggest limitation is that each wrapper is an IE system customized for a particular site and is not universally applicable. Also, source documents are designed for people and few sites provide machine readable specifications of their formatting conventions. Here, ad hoc formatting conventions, used in one site, are rarely relevant elsewhere. Harvest and FAQ Finder, discussed in Section 2, also have two key limitations. First, both systems focus exclusively on web documents and ignore services for e.g., web-log analysis, performance analysis and customer relationship management. Second, both rely on a pre-specified description of certain fixed classes of web documents.

3.3 Generalization

The following difficulties may arise during this task:

Clustering: IR community has explored document clustering as an alternative method of organizing retrieved results, but clustering is yet to be deployed on the major search engines. Google [5], which seems to be the most effective search engine till date, currently supports simple hostname based clustering. Besides, there are some problems in efficient clustering arising out of the nature of web data itself. As mentioned in Section 2, the data is not only distributed, heterogeneous and imprecise, it is also very high dimensional and overlapping. Thus, existing conventional clustering techniques find difficulty in handling these characteristics.

Outliers: The web server, which logs the data of all users and of their transactions, has many outliers (bad observations), including incomplete/noisy/vague data due to various reasons inherent in web browsing and logging. These outliers are not a very small percentage of the database since many users just follow links, which are easily visible, big and prominent. These outliers, in web log server data during WUM, mainly arise because users end up traversing paths which are not in accordance with their interests. Since information on the web is distributed spotting outliers is difficult without clustering the data.

Association Rule Mining: In association rule mining the current techniques are not able to appropriately mine for linguistic association rules which are more human understandable. Some algorithms which convert linguistic rules to numeric ones suffer from the problem of 'hard' rejection. Also, the use of sharp boundary intervals is not intuitive with respect to human perception. For example, an interval method may classify a person as young if the age is less than 35, and old if it is greater than 35 years. This obviously does not always correspond to the human perception of "young" and "old" which considers the boundaries of these imprecise concepts, not hard/crisp.

3.4 Analysis

The biggest problem faced in this step is from the point of view of knowledge discovery and modelling. Discovering knowledge out of the information available on the web has always been a challenge to the analysts, as the output of knowledge mining algorithms is often not suitable for direct human interpretation. This is so, because the patterns discovered are mainly in mathematical form.

4 Soft Computing and its Relevance

Soft computing is a consortium of methodologies which work synergestically and provides in one form or another flexible information processing capabilities for handling real life ambiguous situations. Its aim is to exploit the tolerance for imprecision, uncertainty, approximate reasoning and partial truth in order to achieve *tractability, robustness, low cost solutions,* and *close resemblance to human like decision making* [76]. In other words, it provides the foundation for the conception and design of high MIQ (Machine IQ) systems, and therefore forms the basis of future generation computing systems. At this juncture, Fuzzy Logic (FL), Rough Sets (RS), Artificial Neural Networks (ANN) and Genetic Algorithms (GA) are the principal components where FL provides algorithms for dealing with imprecision and uncertainty arising from vagueness rather than randomness, RS for handling uncertainty arising from limited discernibility of objects, ANN the machinery for learning and adaptation, and GA for optimization and searching.

Relevance of soft computing to pattern recognition and image processing is extensively established in literature [55] [56]. Recently the application of soft computing to data mining problems has also drawn the attention of researchers. A recent review [68] is a testimony in this regard. Here fuzzy logic is used for handling issues related to incomplete/imprecise data/query, approximate solution, human interaction (linguistic information), understandability of patterns and deduction, and mixed media information (fusion). Neural networks are used for modelling highly nonlinear decision boundaries, generalization and learning (adaptivity), self organization, rule generation and pattern discovery. Genetic algorithms are seen to be useful for prediction and description, efficient search, and adaptive and evolutionary optimization of complex objective functions in dynamic environments. Rough set theory is used to obtain approximate description of objects in a granular universe in terms of its *core* attributes. It provides 'fast' algorithms for extraction of domain knowledge in the form of logical rules. Recently various combinations of these tools have been made in soft computing paradigm, among which neurofuzzy integration is the most visible one [56]. In this context we mention about the computational theory of perception which is explained recently by Zadeh [77] as the basic theory behind performing the tasks like driving a car in a city, cooking a meal and summarizing a story, in our day to day life. Here computation may be done with perception which is fuzzy- granular in nature.

Web data, being inherently unlabelled, imprecise/incomplete, heterogeneous and dynamic, appears to be a very good candidate for its mining in soft computing framework. Besides, human interaction being a key component in web mining, as mentioned in Section 2, issues such as context sensitive and approximate queries, summarization and deduction, and personalization and learning are of utmost importance where soft computing seems to be the most appropriate paradigm for providing effective solutions. This realization has drawn the attention of soft computing community to develop 'soft web mining' systems [59] in parallel to the conventional ones since its inception in or around 1996. In the following sections we discuss some such applications of each of the soft computing tools.

5 Fuzzy Logic for Web Mining

The application of fuzzy logic, so far made, to web mining tasks mainly falls under IR and generalization (clustering, association). These attempts will be described here along with different commercially available systems. Some of the prospective areas which need immediate attention are also outlined.

5.1 Information Retrieval

Yager describes in [73] a framework for formulating linguistic and hierarchical queries. It describes an IR language which enables users to specify the inter-relationships between desired attributes of documents sought using linguistic quantifiers. Examples of linguistic quantifiers include 'most', 'at least', 'about half'. Let Q be a linguistic expression corresponding to a quantifier such as 'most' then it is represented as a fuzzy subset Q over $I = [0, 1]$ in which, for any proportion r, belonging to I, $Q(r)$ indicates the degree to which r satisfies the concept indicated by the quantifier Q. Koczky and Gedeon [22] deal with the problem of automatic indexing and retrieval of documents where it cannot be guaranteed that the user queries include the actual words that occur in the documents that should be retrieved. Fuzzy tolerance and similarity relations are presented and the notion of 'hierarchical co-occurrence' is defined that allows the introduction of two or more hierarchical categories of words in the documents.

5.2 Generalization

Clustering: Etzioni [17] has listed the key requirements of web document clustering as measure of relevance, browsable summaries, ability to handle overlapping data, snippet tolerance, speed and incremental characteristics. In [38] fuzzy c medoids (FCNdd) and fuzzy c Trimmed medoids (FCTMdd) are used for clustering of web documents and snippets (outliers). In [28] a fuzzy clustering technique for web log data mining is described. Here, an algorithm called CARD (competitive agglomeration of relational data) for clustering user sessions is described, which considers the structure of the site and the URL's for computing the similarity between two user sessions. This approach requires the definition and computation of dissimilarity/similarity between all session pairs, forming a similarity or fuzzy relation matrix, prior to clustering. Since the data in a web session involves access method (GET/ POST), URL, transmission protocol (HTTP/ FTP) etc. which are all non numeric; correlation between two user sessions and hence their clustering is best handled using fuzzy set approach. Other techniques for clustering web data include those using hypergraph based clustering [51].

Association Rule Mining: Some algorithms for mining association rules using fuzzy logic techniques have been suggested in [26]. They deal with the problem of mining fuzzy association rules understandable to humans from a database containing both quantitative and categorical attributes. Association rules of the form, if X is A, then Y is B where X, Y are attributes and A, B are fuzzy sets, are mined. Nauck [52] has developed a learning algorithm that creates *mixed* fuzzy rules involving both categorical and numerical attributes.

5.3 Prospective Areas of Application

Search Engines: There is immense scope of applying fuzzy logic to improve web search from the points of view of deduction, matching and ranking, among others. To add human like deductive capability to search engine, the use of fuzzy logic is not an option, rather it is a necessity. Regarding matching, a probable approach is to compromise sightly on precision (which is anyway very difficult to achieve due to millions of web pages), and retrieve most 'relevant' documents from an expanded domain. The retrieved documents may then be clustered during/after search, or filtered at the client side, or both. The concept of linguistic variables and membership functions can be used for keyword matching. Similarly, for page ranking the degree of closeness of hits in a document can be used for its computation. For example, variables like 'close', 'far', 'nearby' may be used to represent the distance between hits in a document for a given query. Similarly fuzzy variables like 'reputation', 'importance', attached to the URL which is referencing a particular page, can be used in calculating page ranks. For example, in Fig. 5, which shows a neuro-fuzzy IR system, match parameters such as 'proximity' and subjectivity in queries can be found using fuzzy sets.

Let us consider the popular Google search engine which is considered highly effective among the existing ones. In [5] a schematic diagram (Fig. 2) of the technology behind Google has been explained in which we can see that the lexicon gives wordIDs to each word of the query and wordIDs are then matched. If the query contains quantifiers like less, very less, more, then instead of blindly rejecting/ selecting pages based on their absence/ presence in the document, a smoother transition based on their membership value is a better option. Considering fuzzy queries i.e., if the query text includes linguistic variables like almost, somewhat, more or less, about, we can provide more relevant documents by giving grades of membership to different results. When we consider hits, greater weight should be given to documents in which query words are closer to each other than those in which they are far apart.

Similarity Measures: There are certain questions like: What is the distance between two URLs? Which two URL's are always requested together? Which users have common interests and request similar documents? that appear to be better handled in a fuzzy set theoretic framework since answers to these questions need not always be crisp.

Others: Some other areas where fuzzy logic may be applied include:

- Ontology
- Matching techniques

- Recognition technology
- Summarization
- E-commerce
- Content management
- Database querying
- Information aggregation and fusion
- Customization and profiling

6 Neural Networks and Learning Systems for Web Mining

A neural network can formally be defined as: *a massively parallel interconnected network of simple (usually adaptive) processing elements which is intended to interact with the objects of the real world in the same way as biological systems do.* NNs are designated by the network topology, connection strength between pairs of neurons (called weights), node characteristics and the status updating rules. Normally an objective function is defined which represents the complete status of the network and the set of minima of it corresponds to the set of stable states of the network. Neural network based systems are usually reputed to enjoy the following major characteristics: generalization capability, adaptivity to new data/information, speed due to massively parallel architecture, robustness to missing, confusing, ill-defined/noisy data, and capability for modelling nonlinear decision boundaries.

Neural networks have been applied, so far, to the tasks like IR, IE and clustering (self organization) of web mining, and for personalization. We summarize the existing literature on these lines as follows. Some of the prospective areas which need immediate attention are also discussed.

6.1 Information Retrieval

Artificial neural networks provide a convenient method of knowledge representation for IR applications. Also their learning ability helps to achieve the goal of implementing adaptive systems. Shavlik [65] suggests an agent, the WAWA- IE+IR system (Wisconsin Adaptive Web Assistant), using neural networks with reinforcement learning, which uses two network modules, namely, ScorePage and ScoreLink. ScoreLink uses unsupervised learning while ScorePage uses supervised learning in the form of advice from the users. The system uses KBNN (Knowledge based neural networks) as its knowledge base to encode the initial knowledge of users which is then refined. This has the following advantages: (a) the agent is able to perform reasonably well initially because it can utilize the users' prior knowledge, and (b) users' prior knowledge does not have to be correct as it is refined through learning. Information is derived by extracting rules from KBNN's [66]. In order to map large sized web pages into fixed sized neural networks, a concept of sliding window is used. This parses each page considering three words at a time, and the html tags like $< p >, < /p >, < br >$ act as window breakers. Using self generated training examples it can act also as a self tuning agent. Rules of the type: when *"precondition"* then *"action"* are extracted where actions could be of the type: *strength* followed by 'show page' or 'follow link' or 'avoid showing page'. Here, *strength* could be *weakly, moderately, strongly* or *definitely* which are determined by the weight of the links between layers of the neural network.

Mercure [4] is an example of an IR system, based on multi layered networks, that allows document retrieval using a spreading activation process and query optimization using relevance back-propagation. This model consists of an input layer which represents users information needs, a term neuron layer, a document neuron layer, and an output layer representing the result of query evaluation.

Lim [41] has developed a concept of visual keywords which are abstracted and extracted from visual documents using soft computing techniques. Each visual keyword is represented as a neural network or a soft cluster center. Merkl and Rauber [46] have shown how to use hierarchical feature maps to represent the contents of a document archive. After producing a map of the document space using self-organizing maps, the system provides a hierarchical view of the underlying data collection in the form of an atlas. Using such a modelling the user can easily 'zoom' into particular regions of interest while still having general maps for overall orientation. A ANN based hybrid web text mining system has been described in [21].

6.2 Information Extraction

Most IE systems that use learning fall into two groups: the one that uses relational learning [20], [69] to learn 'extracted patterns', and the other group learns parameters of Hidden Markovs Models (HMM) and uses them to extract information [2]. In [19] wrapper induction techniques are combined with adaBoost algorithm (Schapire and Singer, 1998) called Boosted Wrapper Induction (BWI) and the system has outperformed many of the relational learners and is competitive with WAWA-IE and HMM. For a brief and comprehensive view of the various learning systems used in web content mining, one may refer to [36].

6.3 Self-Organization (WEBSOM)

The emerging field of text mining applies methods of data mining and exploratory data analysis to analyze text collections and to convey information to the users in an intuitive manner. Visual map-like displays provide a powerful and fast medium for portraying information about large collections of text. Relationships between text items and collections, such as similarity, clusters, gaps and outliers, can be communicated naturally using spatial relationships, shading and colors. In WEBSOM [35] the self-organizing map (SOM) algorithm is used to automatically organize very large and high-dimensional collections of text documents onto two-dimensional map displays. The map forms a document landscape where similar documents appear close to each other at different points of the regular map grid. The landscape can be labelled with automatically identified descriptive words that convey properties of each area and also act as landmarks during exploration. With the help of an HTML-based interactive tool the ordered landscape can be used in browsing the document collection and in performing searches on the map. An organized map offers an overview of an unknown document collection helping the user in familiarizing oneself with the domain. Map displays that are already familiar can be used as visual frames of reference for conveying properties of unknown text items. Thematically arranged static document landscapes provide meaningful background for dynamic visualizations of time-related properties of the data, for example. The mathematical preliminaries, background, basic ideas, implications, and numerous applications of self-organizing maps are described in a recent book [34].

6.4 Personalization

Personalization means that the content and search results are tailored as per users interests and habits. Neural networks may be used for learning user profiles with training data collected from users or systems as in [65]. Since user profiles are highly non-linear functions, neural networks seem to be an effective tool to learn them. An agent which learns user profiles using Bayesian classifier is "Syskill & Webert" [61]. Once the user profiles have been learned, it can be used to determine whether the users would be interested in another page. However, this decision is made by analyzing the HTML source of a page, and it requires the page to be retrieved first. To avoid network delays, we allow the user to prefetch all pages accessible from the index page and store them locally. Once this is done, Syskill & Webert can learn a new profile and make suggestions about pages to visit quickly. Once the HTML is analyzed, it annotates each link on the page with an icon indicating the user's rating or its prediction of the user's rating together with the estimated probability that a user would like the page. Note that these ratings and predictions are specific to only one user and do not reflect on how other users might rate the pages. As described above, the agent is limited to making suggestions about which link to follow from a single page only. This is useful when someone has collected a nearly comprehensive set of links about a topic. A similar system which assists users in browsing software libraries has been built by Drummond [13].

6.5 Prospective Areas of Application

Personalized Page Ranking: As mentioned in Section 2.1, page ranks are important since human beings find it difficult to scan through the entire list of documents returned by the search engine in response to his/her query. Therefore, it is desirable, for convenience, to get the pages ranked with respect to 'relevance' to user queries so that one can get the desired documents only by scanning the first few pages.

Let us consider here again the case of the popular search engine Google [5]. It computes the rank of a page *'a'* using:

$$\Pr(a) = 1 - d + d \sum_{i=1}^{n} \frac{\Pr(T_i)}{C(T_i)} \tag{1}$$

where d is the damping factor, $\Pr(a)$ is the rank of page *'a'* which has pages $T_1, T_2, \ldots, T_n$ pointing to it and $C(a)$ is the number of outgoing links from page *'a'*.

Note that it takes into consideration only the popularity of a page (reputation of incoming links) and richness of information content (number of outgoing links) and does not take care of other important factors like:

- User preference: Whether the link matches with the 'preferences' of the user, established from his/ her history ?
- Validity: Whether the link is currently valid or not?
- Interestingness: Whether the page is of overall 'interest' to the user or not?

These should also be reflected in the computation of page ranks. The learning/ generalization capability of ANN's can be exploited for determining user preference and interestingness. user preference can be incorporated by training a neural network based on user history. Since ANN's can model non linear functions and learn from examples, they appear to be a good candidate for computing the 'interestingness' of a page. Self organizing neural networks can be used to filter out invalid pages dynamically.

An ANN can compute the page rank from a combination of each of the parameters like hub, authority, reputation, validity, interestingness, user preference with weights assigned to each which the user can modify; thereby refining the network as per his personalized interest. These factors may sometimes also be characterized by fuzzy variables. For example, variables like 'close', 'far', 'nearby' may be used to represent the distance between hits in a document for a given query. Similarly fuzzy variables like 'reputation', 'importance' can be attached to the URL which is referencing a particular page. This is unlike in Google where these variables are considered to be crisp. This signifies the importance of integrating synergistically ANN with fuzzy logic under *neuro-fuzzy paradigm* [47] for computing page ranks. In the next paragraph we describe more details of a proposed neuro-fuzzy IR system.

Neuro-fuzzy IR: A schematic diagram of a proposed neuro-fuzzy IR system is shown in Fig. 5. It shows that the total 'relevance' of a document is the

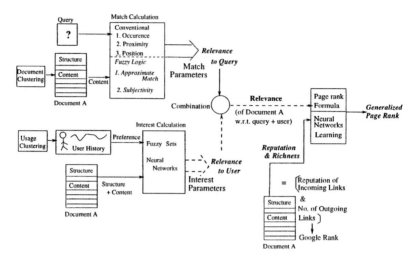

Fig. 5. Neuro fuzzy IR system

combination of 'relevance with respect to a query' (match parameters) and 'relevance with respect to the user' (interest parameters). This total relevance, when combined with 'richness and reputation' of document A (currently reflected in Google rank) give the 'generalized' page rank. The dotted links represent areas not addressed by existing algorithms. Means of computing each of the quantities, namely, 'relevance with respect to a query', 'relevance with respect to the user' and 'richness and reputation', are mentioned below:

Relevance to a query: Here fuzzy logic can be used in computing match parameters by handling subjectivity in queries and approximate matching in documents; thereby better modelling 'relevance to query'. Also structured documents can be handled more effectively in the framework of fuzzy logic. Literature described in Section 5.1 address many of these tasks.

Relevance with respect to a user: Literature in this area is relatively scarce. Existing approaches belong mainly to three categories: (a) learning from user 'history' or 'profile', (b) clustering of users into homogeneous groups, and (c) using relevance feedback. ANN's can be used to learn the non-linear user profiles from their previous history and reflect 'relevance to user' of a document A in interest parameters.

Richness and reputation: This parameter is reflected in most existing page ranking systems. However, efficient computation of the page rank is an open research issue where neural networks may be used.

Clustering and Classification: Neural networks can be used to classify web pages as well as user patterns, in both supervised and unsupervised modes. Its ability in modelling complex nonlinear functions can also be exploited here.

Deduction: Another area where neural networks may be used is in building deductive capabilities in web mining systems. As mentioned earlier complex,

250

nonlinear functions may be learned using neural networks and logical rules may be extracted from trained networks using rule extraction algorithms. The logical rules are human interpretable and help in generating deductions.

7 Genetic Algorithms for Web Mining

Genetic algorithms (GAs), a biologically inspired technology, are randomized search and optimization techniques guided by the principles of evolution and natural genetics. They are efficient, adaptive and robust search processes, producing near optimal solutions and have a large amount of implicit parallelism. GAs are executed iteratively on a set of coded solutions (genes), called population, with three basic operators: selection/reproduction, crossover and mutation. They use only the payoff (fitness function) information and probabilistic transition rules for moving to the next iteration.

The literature explaining the use of genetic algorithms to web mining seems to be even poorer than that of fuzzy logic and neural networks. Genetic algorithms are used, mainly in search, optimization and description. Here we describe some of the attempts:

Search and Retrieval: A GA-based search to find other relevant homepages, given some user supplied homepages, has been implemented in G-Search [11]. Web document retrieval by genetic learning of importance factors of HTML tags has been described in [32]. Here, the method learns the importance of tags from a training text set. Genetic operators for documents searching has been defined in [62]. GAs have also been applied for the purpose of feature selection in text mining [45].

Query Optimization: In [3], Boughanem et. al. have developed a query reformulation technique using genetic algorithms in which a GA generates several queries that explore different areas of the document space and determines the optimal one. Yang et.al. [75] have presented an evolutionary algorithm for query optimization by re-weighting the document indexing without query expansion. Kraft et.al. [37] apply genetic programming in order to improve weighted boolean query formulation.

Document Representation: Gordon [25] adopted a GA to derive better descriptions of documents. Here each document is assigned N descriptions where each description is a set of indexing terms. Then genetic operators and relevance judgements are applied to these descriptions in order to determine the best one in terms of classification performance in response to a specific query. Automatic web page categorization and updating can also be performed using GAs [43].

Distributed Mining: GEMGA (Gene Expression Messy Genetic Algorithm) [29] which is a sub-quadratic, highly parallel evolutionary search algorithm is specially found suitable for distributed data mining applications including the web. Foundation of GEMGA is laid on the principle of both decomposing

251

black box search into iterative construction of partial ordering and performing selection operation in the relation, class and sample spaces. The GEMGA is designed based on an alternate perspective of evolutionary computation proposed by the SEARCH framework that emphasizes the role of gene expression or intra-cellular flow of genetic information. The research on the use of GEMGA in distributed data mining is growing fast and it deals with the problems of finding patterns in data in an environment where both the data and the computational resource are distributed. In [30], Kar Gupta et. al., suggest CDM (Collective Data Mining) as a new approach towards distributed data mining (DDM) from heterogeneous sites. It is pointed out that naive approaches to distributed data analysis in an heterogeneous environment may lead to an ambiguous situation. CDM processes information locally and builds a global model in situations where it is not possible, as in the case of web, to collect all the data in a common data warehouse and then process.

Web Spiders: Web spiders are web agents that automate data retrieval and processing. A Web 'spider' is a type of 'bot', 'crawler' or robot/automated software which explores the World Wide Web, retrieving a document and following all the hyperlinks in it. The program then can generate a catalog which can be accessed by search engines. With the growth of internet in complexity and size, spiders are increasingly becoming specialized and difficult to code. Genetic algorithms, with their capability for evolution and efficient optimization, provide a promising tool for web spider design. One such intelligent architecture for automatic design of agents is described in [8].

Regarding the prospective areas of application of GAs, let us consider here the case of *Adaptive Web Sites.* These are sites which automatically improve their organization and presentation by learning from visitor access patterns [15]. It focuses on the problem of *index page synthesis* where an index page is a page consisting of a set of links that cover a particular topic. Its basic approach is to analyze web access logs and find groups of pages that often occur together in user visits (and hence represent coherent topics in users' minds) and to convert them into 'index' pages. Here genetic algorithms may be used for prediction of user preferences, dynamic optimization and evolution of web pages.

8 Rough Sets for Web Mining

Rough sets are characterized by their ability for granular computation. In rough set theory a concept B is described by its 'lower' ($\underline{B}$) and 'upper' ($\bar{B}$) approximations defined with respect to some indiscernibility relation. The use of rough set theory for knowledge discovery [67] and rule mining [57, 1] is widely acknowledged. However, the current literature on application of rough sets to web mining, like genetic approach, is very scanty. Some web mining tasks where rough set theory have been used are mentioned below.

8.1 Information Retrieval

Granular Information Retrieval: An important application of rough sets is in 'granular information retrieval'. In rough set theory *Information granules* refer to homogeneous blocks/ clusters of documents as described by a particular set of features which may vary over the clusters. The approach is efficient in many document retrieval tasks where the documents are clustered into homogeneous groups before they are retrieved. Documents are often represented by a large number of features or words and dimensionality reduction needs to be performed for clustering. However, instead of representing all the clusters by the same set of features (words), in granular IR using rough set theory each cluster is represented by different feature sets. This is closer to reality because different word sets are important for different document classes. Wong et. al [72] suggests reducing the dimensionality of terms by constructing a term hierarchy in parallel to a document hierarchy.

Handling Heterogeneous Data: Rough set as well as rough-fuzzy hybrid systems [58] have been used for handling multimedia data and information fusion. A system where rough sets have been used for retrieval of multimedia objects is described in [70].

8.2 Association/ Clustering

Rough sets have been used for document clustering [31] and mining of web usage patterns [44]. Uses of variable precision rough sets [44] and tolerance relations are important in this regard.

Some additional areas where rough sets may be applied include :

- Web Structure Mining: Rough mereology [64], which has ability for handling complex objects, is a potential tool for mining multimedia objects as well as complex representations like web graphs, semantic structures.
- Multi agent systems and collaborative mining.
- Rough-neuro computing (RNC), as a means of computing with words (CWW), is likely to play an important role in natural language query formulation.
- Rough set theory can also be used for the purpose of approximate information retrieval, where the set of relevant documents may be rough and represented by its 'upper' and 'lower' approximations. The lower approximation refers to the set which is definitely relevant and the upper approximation denotes the maximal set which may be possibly relevant. Dynamic and focussed search, exploiting the above concept, may also help in developing efficient IR systems.

9 Conclusions and Discussion

In this chapter, we have summarized the different types of web mining and its basic components, along with their current states of art. The limitations of the existing web mining methods/tools are explained. The relevance of soft computing, including integration of its constituting tools, is illustrated through examples and diagrams. Their applications to each web mining task along with the commercially available systems are described. Lastly, the possible future directions of using fuzzy logic, ANN's, GAs and rough sets for some of these tasks are given, in detail.

In addition to those discussed here, some aspects of web mining, in general, where soft computing is likely to play a key role, in future, are as follows:
(a) At present web content is mainly text-centric, and most mining algorithms are oriented towards and developed from text mining framework. However, web is increasingly gaining a multimedia character with pages containing images, videos etc. Web mining algorithms having capabilities for handling multimedia data need to be developed in near future. Some attempts in this direction are described in [71, 74].
(b) Currently queries are in the form of keywords, advanced search engines may support visual queries. In this regard, the research on content based image retrieval (CBIR) in soft computing framework has potential significance.
(c) Most search engines perform search on English text only, not across languages. With these becoming increasingly common, multilingual search engines and IR systems which can identify languages, translate, perform thematic classification, and can provide summaries automatically are recently being developed. Soft computing may be used to increase the efficiency of such systems.
(d) Collaborative mining and automatic interaction among sites and constitute a recent research area (e.g., the .NET paradigm of Microsoft). Here web query is not the only means to obtain required documents, and answers to queries can be automatically obtained from distributed web resources. Thus text sources could be used for planning and problem solving tasks (e.g., an agent on the web could be used to make ones' travel plans automatically). Significance of applying soft computing for the above tasks may therefore be explored.
(e) Some tasks related to embedded internet systems that could be handled using soft computing include, access control, task scheduling, system configuration, priority order, device monitoring, bug and failure reporting, as well as distributed (remote) control of electronic products (devices).
(f) E-commerce is an important application area of soft computing. It may be used to impart human like interaction in e-shopping portals [39]. For example, buyers interest can be better modelled (as functions of price and quality) using fuzzy set theory.
(g) The development of new knowledge visualization techniques for effective user interface may also be done with soft computing.

Finally, (soft) case based reasoning (CBR) [54], which is a popular AI problem solving paradigm, using soft computing tools, and is recently drawing the attention of researchers worldwide, may be used for solving many of the web mining problems, stated above. In this context, the mention may be made of the computational theory of perception, recently explained by Zadeh [77], which is characterized mainly using the concept of "fuzzy- granularity" of perceptions.

References

1. M. Banerjee, S. Mitra, and S. K. Pal. Rough fuzzy MLP: Knowledge encoding and classification. *IEEE Transactions on Neural Networks*, 9:1203–1216, 1998.
2. D. Bikel, R. Schwartz, and R. Weischedel. An algorithm that learns what's in a name. *Machine learning*, 34 (Special issue on Natural Language Learning)(1/3):211–231, 1999.
3. M. Boughanem, C. Chrisment, J. Mothe, C. S. Dupuy, and L. Tamine. Connectionist and genetic approaches for information retrieval. In F. Crestani and G. Pasi, editors, *Soft Computing in Information Retrieval: Techniques and Applications*, volume 50, pages 102–121. Physica Verlag, Heidelberg, 2000.
4. M. Boughanem, T. Dkaki, J. Mothe, and C. Soule-Dupuy. Mercure at trec7. In *Proceedings of the 7th International Conference on Text Retrieval, TREC7*, Gaithrsburg, MD, 1998.
5. S. Brin and L. Page. The anatomy of a large scale hypertextual web search engine. In *Proceedings of Eighth International WWW Conference*, pages 107–117, Brisbane, Australia, April 1998.
6. S. Chakrabarti. Data mining for hypertext. *ACM SIGKDD Explorations*, 1(2):1–11, 2000.
7. S. Chakrabarti, M. van den Berg, and B. Dom. Focused crawling: A new approach to topic-specific web resource discovery. In *Proceedings of the 8th World Wide Web Conference*, Toronto, May 1999.
8. H. Chen, Y. Chung, C. Yang, and M. Ramsey. A smart itsy bitsy spider for the web. *Journal of the American Society for Information Science*, 49(7):604–618, 1997.
9. W. W. Cohen. What can we learn from the web? In *Proceedings of 16th International Conference on Machine Learning (ICML99)*, pages 515–521, 1995.
10. R. Cooley, B. Mobasher, and J. Srivastava. Web mining:information and pattern discovery on the world wide web. In *Proceedings of the 9th IEEE International Conference on Tools with Artificial Intelligence*, Newport beach, CA, November 1997.
11. F. Crestani and G. Pasi, editors. *Soft Computing in Information Retrieval: Techniques and Application*, volume 50. Physica-Verlag, Heidelberg, 2000.
12. D. D. Cutting, J. Karger, J. Pederson, and J. Scatter. A cluster based approach to browsing large document collections. *Proceedings of the Fifteenth International Conference on Research and Development in Information Retrieval*, pages 318–329, June 1992.
13. C. Drummond, D. Ionescu, and R. Holte. A learning agent that assists the browsing of software libraries. Technical Report TR-95-12, University of Ottawa, 1995.
14. O. Etzioni. Moving up the information food cahin: Deploying softbots on the web. In *Proceedings of the Fourteenth National Conference on AI*, pages 1322–1326, Portland, OR, 1996.

15. O. Etzioni and M. Perkowitz. Adaptive web sites: An AI challenge. In *Proceedings of Fifteenth National Conference on Artificial Intelligence*, Madison, Wisconsin, July 1998.
16. O. Etzioni, J. Shakes, and M. Langheinrich. Ahoy! the homepage finder. In *Proceedings of Sixth WWW Conference*, Santa Carla, CA, April 1997.
17. O. Etzioni and O. Zamir. Web document clustering: A feasibility demonstration. In *Proceedings of the 21st Annual International ACM SIGIR Conference*, pages 46-54, 1998.
18. D. Freitag. Information extraction from html: Application of a general machine learning approach. In *Proceeding of Fifteenth Conference on Artificial Intelligence AAAAI-98*, pages 517-523, 1998.
19. D. Freitag and N. Kushmerick. Boosted wrapper induction. In *Proceedings of AAAI*, pages 577-583, 2000.
20. D. Freitag and A. McCallum. Information extraction from hmm's and shrinkage. In *Proceedings of AAAI-99 Workshop on Machine Learning for Information Extraction*, Orlando, FL, 1999.
21. H. Fukuda, E.L.P. Passos, A. M. Pacheco, L. B. Neto, J. Valerio, V. Jr. De Roberto, E. R. Antonio, and L. Chigener. Web text mining using a hybrid system. In *Proceedings of the Sixth Brazilian Symposium on Neural Networks*, pages 131-136, 2000.
22. T. Gedeon and L. Koczy. A model of intelligent information retrieval using fuzzy tolerance relations based on hierarchical co-occurrence of words. In F. Crestani and G. Pasi, editors, *Soft Computing in Information Retrieval: Techniques and Applications*, volume 50, pages 48-74. Physica Verlag, Heidelberg, 2000.
23. R. Ghani, R. Jones, D. Mladenic, K. Nigam, and S. Slattery. Data mining on symbolic knowledge extracted from the web. In *Proceedings of the Sixth International Conference on Knowledge Discovery and Data Mining (KDD-2000) Workshop on Text Mining Boston, MA*, pages 29-36, August 2000.
24. D. Gibson. Inferring web communities from link topologies. In *UK conference on Hypertext*, 1998.
25. M. D. Gordon. Probabilistic and genetic algorithms for document retrieval. *Communications of the ACM*, 31(10):208-218, 1988.
26. A. Gyenesei. A fuzzy approach for mining quantitative association rules. TUCS technical reports 336, University of turku, Department of Computer Science, Lemminkisenkatu14, Finland, March 2000.
27. J. Hipp, U. Guntzer, and J. Nakhaeizadeh. Algorithms for association rule mining- a general survey and comparison. *ACM SIGKDD Eplorations*, 2(1):58-65, July 2000.
28. A. Joshi and R. Krishnapuram. Robust fuzzy clustering methods to support web mining. In *Proc Workshop in Data Mining and Knowledge Discovery, SIGMOD*, pages 15-1 to 15-8, 1998.
29. H. Kargupta. The gene expression messy genetic algorithm. In *Proceedings of the IEEE International Conference on Evolutionary Computation*, pages 631-636, Nagoya University, Japan, 1996.
30. H. Kargupta, B. H. Park, D. Hershberger, and E. Johnson. Collective data mining: A new perspective toward distributed data mining. *Advances in Distributed and Parallel Knowledge Discovery*, 1999. MIT/AAAI Press.
31. S. Kawasaki, N. Binh Nguyen, and T. Bao Ho. Hierarchical document clustering based on tolerance rough set model. In *Proceedings of the Sixth International*

Conference on Knowledge Discovery and Data Mining (KDD-2000) Workshop on Text Mining Boston, MA, August 2000.

32. S. Kim and B. Thak Zhang. Web document retrieval by genetic learning of importance factors for html tags. In *Proceedings of the International Workshop on Text and Web mining*, pages 13–23, Melbourne, Australia, August 2000.

33. Jon M. Kleinberg. Authoritative sources in a hyperlinked environment. *Journal of the ACM*, 46(5):604–632, 1999.

34. T. Kohonen. *Self-organising Maps*. Springer, Berlin, Germany, second edition, 1997.

35. T. Kohonen. Self organizing maps for large documents. *IEEE Transactions on Neural networks*, 11 (Special issue on Data Mining)(3):574–589, June 2000.

36. R. Kosla and H. Blockeel. Web mining research:a survey. *SIG KDD Explorations*, 2:1–15, July 2000.

37. D. H. Kraft, F. E. Petry, B. P. Buckles, and T. Sadasivan. The use of genetic programming to build queries for information retrieval. In *Proceedings of the IEEE Symposium on Evolutionary Computation*, Orlando, FL, 1994.

38. R. Krishnapuram, A. Joshi, and L. Yi. A fuzzy relative of the k-medoids algorithm with application to document and snippet clustering. In *Proceedings of IEEE Intl. Conf. Fuzzy Systems - FUZZIEEE 99, Korea*, 1999.

39. C.-H. Lee and H.-C. Yang. Developing an adaptive search engine for e-commerce using a web mining approach. In *Proceedings of the International Conference on Information Technology: Coding and Computing*, pages 604–608, 2001.

40. A.Y. Levy and D.S. Weld. Intelligent internet systems. *Artificial Intelligence*, 118(1-2), 2000.

41. J. H. Lim. Visual keywords: from text retrieval to multimedia retrieval. In F. Crestani and G. Pasi, editors, *Soft Computing in Information Retrieval: Techniques and Applications*, volume 50, pages 77–101. Physica Verlag, Heidelberg, 2000.

42. W. Y. Lin, S. A. Alvarez, and C. Ruiz. Collaborative recommendation via adaptive association rule mining, August 2000.

43. V. Loia and P. Luongo. An evolutionary approach to automatic web page categorization and updating. In N. Zhong, Y. Yao, J. Liu, and S. Oshuga, editors, *Web Intelligence: Research and Developement*, volume LNCS 2198, pages 292–302. Springer Verlag, Singapore, 2001.

44. V. Uma Maheswari, A. Siromoney, and K. M. Mehata. The variable precision rough set model for web usage mining. In *Proceedings of the First Asia-Pacific Conference on Web Intelligence (WI-2001)*, Maebashi, Japan, October 2001.

45. M.J. Martin-Bautista and M.-A. Vila. A survey of genetic feature selection in mining issues. In *Proceedings of the Congress on Evolutionary Computation (CEC 99)*, pages 13–23, 1999.

46. D. Merkl and A. Rauber. Document classification with unsupervised artificial neural networks. In F. Crestani and G. Pasi, editors, *Soft Computing in Information Retrieval: Techniques and Applications*, volume 50, pages 102–121. Physica Verlag, Heidelberg, 2000.

47. S. Mitra and S. K. Pal. Fuzzy multi-layer perceptron, inferencing and rule generation. *IEEE Transactions on Neural Networks*, 6:51–63, 1995.

48. D. Mladenic and M. Grobelnik. Efficient text categorization. In *Proceedings of Text Mining Workshop on the 10th European Conference on Machine Learning ECML98*, 1998.

49. B. Mobasher, H. Dai, T. Luo, M. Nakagawa, Y. Sun, and J. Wiltshire. Discovery of aggregate usage profiles for web personalization. In *Proceedings of KDD-2000 Workshop on Web Mining for E-Commerce*, Boston, MA, August 2000.

50. B. Mobasher, N. Jain, E-Hong(Sam) Han, and J. Srivastava. Web mining: Patterns from from WWW transactions. Technical Report TR96-050, Department of Computer Science,University of Minnesota, March 1997.

51. B. Mobasher, V. Kumar, and E. H. Han. Clustering in a high dimensional space using hypergraph models. Technical Report TR-97-063, University of Minnesota, Minneapolis, 1997.

52. D. Nauck. Using symbolic data in neuro-fuzy classification. In *Proceedings of NAFIPS'99, New York, USA*, pages 536–540, June 1999.

53. C. V. Negotia. On the notion of relevance in information retrieval. *Kybernetes*, 2(3):161–165, 1973.

54. S. K. Pal, T. S. Dillon, and D. S. Yeung, editors. *Soft computing in Case Based Reasoning*. Springer-Verlag, London, 2000.

55. S. K. Pal, A. Ghosh, and M. K. Kundu, editors. *Soft Computing for Image Processing*. Physica Verlag, Heidelberg, 2000.

56. S. K. Pal and S. Mitra. *Neuro-Fuzzy Pattern Recognition: Methods in Soft Computing*. John Wiley, New York, 1999.

57. S. K. Pal, S. Mitra, and P. Mitra. Rough fuzzy MLP: Modular evolution, rule generation and evaluation. *IEEE Transactions on Knowledge and Data Engineering*, to appear, 2001.

58. S. K. Pal and A. Skowron. *Rough Fuzzy Hybridization: A New Trend in Decision Making*. Springer-Verlag, Singapore, 1999.

59. S. K. Pal, V. Talwar, and P. Mitra. Web mining in soft computing framework: Relevance, state of the art and future directions. *IEEE Trans. Neural Networks*, 13(5):1163–1177, 2002.

60. J. Pazzani and D. Billsus. Learning collaborative information filters. In *Proceedings of Fifteenth International Conference on Machine Learning*, Madison, Wisc, 1998. Morgan Kauffman.

61. M. Pazzani, J. Muramatsu, and D. Billsus. Syskill and webert:identifying interesting web sites. In *Proceedings of Thirteenth National Conference on AI*, pages 54–61, 1996.

62. F. Picarougne, N. Monmarche, A. Oliver, and G. Venturini. Web mining with a genetic algorithm. In *Proceedings of the Eleventh International World Wide Web Conference*, Hawaii, 2002.

63. J. Pitkow. In search of reliable usage data on the www. In *Proceedings of the Sixth International WWW conference*, pages 451–463, Santa Carla, CA, 1997.

64. L. Polkowski and A. Skowron. Rough mereology: A new paradigm for approximate reasoning. *International Journal of Approximate Reasoning*, 15(4):333–365, 1996.

65. J. Shavlik and T. Eliassi. A system for building intelligent agents that learn to retrieve and extract information. *International Journal on User Modeling and user adapted interaction*, April (Special issue on User Modeling and Intelligent Agents 2001.

66. J. Shavlik and G. G. Towell. Knowledge-based artificial neural networks. *Artificial Intelligence*, 70(1/2):119–165, 1994.

67. A. Skowron and L. Polkowski, editors. *Rough Sets in Knowledge Discovery*. Physica-Verlag, Heidelberg, 1998.

68. S.Mitra, S. K. Pal, and P. Mitra. Data mining in soft computing framework: A survey. *IEEE Transactions on Neural Networks*, 13(1):3–14, 2002.
69. S. Soderland. Learning information extraction rules for semi-structured and free text. *Machine learning*, 34 (Special issue on Natural Language Learning)(1/3):233–272, 1999.
70. U. Straccia. A framework for the retrieval of multimedia objects based on four-valued fuzzy description logics. In F. Crestani and G. Pasi, editors, *Soft Computing in Information Retrieval: Techniques and Applications*, volume 50, pages 332–357. Physica Verlag, Heidelberg, 2000.
71. C. Wan, M. Liu, and L. Wang. Content-based sound retrieval for web application. In N. Zhong, Y. Yao, J. Liu, and S. Oshuga, editors, *Web Intelligence: Research and Developement*, volume LNCS 2198, pages 389–393. Springer Verlag, Singapore, 2001.
72. S. K. Wong, Y. Y. Yao, and C .J. Butz. Granular information retrieval. In F. Crestani and G. Pasi, editors, *Soft Computing in Information Retrieval: Techniques and Applications*, volume 50, pages 317–331. Physica Verlag, Heidelberg, 2000.
73. R. Yager. A framework for linguistic and hierarchical queries for document retrieval. In F. Crestani and G. Pasi, editors, *Soft Computing in Information Retrieval: Techniques and Applications*, volume 50, pages 3–20. Physica Verlag, Heidelberg, 2000.
74. K. Yanai, M. Shindo, and K. Noshita. A fast image-gathering system on the world wide web using a PC cluster. In N. Zhong, Y. Yao, J. Liu, and S. Oshuga, editors, *Web Intelligence: Research and Developement*, volume LNCS 2198, pages 324–334. Springer Verlag, Singapore, 2001.
75. J. J. Yang and R. Korfhage. Query modification using genetic algorithms in vector space models. TR LIS045/I592001, Department of IS, University of Pittsburg, 1992.
76. L. A. Zadeh. Fuzzy logic, neural networks, and soft computing. *Communications of the ACM*, 37:77–84, 1994.
77. L. A. Zadeh. A new direction in AI: Towards a computational theory of perceptions. *AI magazine*, 22:73–84, 2001.

A decision support tool for web-shopping using Product Category Summarization

Gabriella Pasi[1], Ronald R. Yager[2]

[1] ITC-CNR- via Ampère 56 - 20131 Milano Italy
   e-mail: gabriella.pasi@itim.mi.cnr.it
[2] Institute of Machine Intelligence - Iona College
   New Rochelle, NY 10801
   e-mail :RYager@Iona.edu

Abstract

The increasing use of the World Wide Web for online shopping challenges us with the objective of designing friendly interfaces that help users in their search for relevant products. To this aim, considerable research is done concerning recommender systems, which are defined to automatically give to consumers some purchase recommendations, based on different techniques. In this paper we describe a decision support method, called Product Category Summarization (PCS), which, given a product category, generates qualitative descriptions of classes of products in order to help consumers in making purchasing decisions. This tool is based on a preliminary clustering of a product line (i.e. 27 inch televisions), into price categories such as low end, moderate and high end. Once this partitioning has been obtained, some linguistic summaries are generated to describe the properties of each category with respect to some relevant features. An example of such a summary is "Most TV's in the high price category provide extremely high resolution". With the aid of such information it becomes much easier for consumers to more easily and confidently locate products that are of particular value.

1 Introduction

In recent years, the growth of the World Wide Web has enabled electronic commerce to become a means for offering consumers a useful way to select and buy products. As any activity which deals with an automatic access to a huge quantity of information, a significant problem when buying on the Internet is related to the location of relevant products with respect to consumers' constraints, both quality and price-based. The focus of this paper is on the web-shopping activity and particularly on ways to improve the quality of the information available to customers in making purchase decisions. For a customer, the appealing characteristic of electronic purchase is to have a wide variety of possible choices. Let us think about the problem of selecting a book or a movie: this is usually done on the basis of several criteria, which are strongly related to the users' tastes and subjectivity. The main problem is that, usually, even if the customer's preferences are

261

well defined, it is very difficult to select the possibly relevant products or services among thousands. In order to help users to more easily identify relevant products, the so called recommender systems have been conceived in order to propose to customers a selection of possibly relevant items, by providing them with a personalized access to information [12]. These systems are based on different techniques among which the most used are collaborative filtering techniques and constraint-based filtering techniques [6]. The former exploit the knowledge about the users with similar tastes to those of the active user, based on an analysis of the products already rated by the active user and of the stored users' profiles. The aim of these systems is to predict the preferences of the active user over the unseen alternatives. The latter systems offer to customers an access to the products based on a description of their features; in such a way the products that possibly reflect the customer's preferences can be easily identified. The concept of preference is in this case related to the fulfilling of several criteria whose evaluation qualifies the tastes of the customer.

In this contribution a method is described, called **Product Category Summarization** (PCS), which has the aim of helping the consumers understanding a product line in a way that can help them in their purchasing decisions. After providing a clustering of a product line into a finite number of categories, typically defined by price, this method automatically constructs some user friendly descriptions of the relevant features shared by the majority of the products associated with each category. By these descriptions a consumer can easily get a good understanding of the issues involved choosing a product in this line.

From a formal point of view the PCS method is constituted by three main steps:

1. *the construction of the Product Space*: the primary component of the product space is a list of features that are useful in describing the product. For example some features useful in describing a TV are screen size, resolution, price, quality of sound system, whether it has flat screen and whether it has picture in picture. The selection of the relevant features is of course the responsibility of a product line expert. The second component in the product space is a listing of the available products along with the values for their relevant features. Furthermore, a good product space should include a textual annotation describing the role of each of the more technical features in contributing to good product. For example with respect the TV some indication of the benefit of flat screen would be useful. We should note that this textual annotation is not essential to our method but provides a knowledge base which enables a consumer to more effectively use the information contained in the product space map.

2. *product space segmentation*: here use is made of the primacy of the price feature in purchasing decisions. Specifically using a clustering technique the available products are partitioned, based upon their price, into a number of categories. These categories should correspond to commonly used consumer conceptualizations of pricing categories denoted by linguistic terms such as *low end, moderate* and *high end*. In order to accomplish this we use a hierarchical clustering technique described in this work.

3. *product space description*: here we make use of the concept of linguistic summaries [18,19] to describe features associated with the typical products

available in the different categories. With the aid of these linguistic summaries we are able to provide consumers with information such as *Most of the high end TV's have extremely high resolution* or *Only a few of the moderate price TV's have flat screens.* These linguistic summaries, which make use of fuzzy set technology, provide a user-friendly description of the objects in the different categories with respect to the relevant features. In future versions of this approach we shall include a reasoning capability which will enable the system to locate atypical objects in the different categories and to be able to answer questions posed by consumers such as *What am I getting by buying a high end product that I don't get in the moderate class.*

The paper is organized as follows: in section 2 an overview of the problem of product brokering in electronic commerce is synthesized. In section 3 the phases of product space construction and segmentation are described. In section 4 the phase of generation of product categories descriptions is illustrated, and in section 5 some conclusions are outlined.

2 The problem of product brokering in electronic commerce

The rapid expansion of the World Wide Web is enabling electronic commerce to become an important means for offering consumers a simple way to purchase products and services. One important issue in electronic commerce concerns online shopping, which encompasses the problems of examining the available products and comparing them in order to select those satisfying the customers' needs [6,12]. A major problem associated with electronic shopping is the cost and time spent to find some relevant information about products and services. In fact, a customer visiting an electronic store often faces considerable difficulty in finding the product that s/he most desires, also due to the fact that there is any assistance in this selection phase. The development of tools which assist customers in making their purchase decisions by customizing the e-shopping environment to users individual preferences is desirable for both consumers and vendors [7]. The availability of decision support tools allows to increase the probability of the customer in undertaking a commercial transaction.

As outlined in [7], there is a trade-off between the effort in a decision-making activity and the accuracy of the obtained results. It follows that there is a need for intelligent tools for purchasing assistance, which reduce the effort of taking purchase decisions, also by means of tolerance to imprecision/vagueness in specifying preferences. This observation supports the benefit of developing modes of information presentation and access which offer consumers simpler ways to select services and products, reducing their cognitive efforts.

The decision-based problem in electronic purchase that is considered in this paper is generally called *product brokering*, which is aimed at retrieving information which helps a user in determining what to buy [6]. In this context, the aim of recommender systems is to suggest to customers a personalized selection of products, and to give them information that can help them to identify possibly interesting products [12]. These systems are based on a wide range of techniques, the three main classes of which are the collaborative filtering techniques, the constraint-based techniques and the content-based techniques [6].

Systems based on collaborative filtering techniques use the preferences expressed by a group of consumers to predict the preferences of a so-called "active user". The basic idea is that the active user will prefer the information items that similar-minded people prefer [10]. The preferences related to a set of products are in this case expressed by the consumers as degrees of satisfaction (also called ratings). So the prediction problem consists in determining the ratings of alternatives not yet graded by the the active user. As outlined in [11] the "collaborative decision support" configures as a new type of decision making problem: in the process of deciding the best products to recommend to a consumer, the other consumers, whose purchase decisions are stored in a database, are seen as "passive" advisors. Differently from group decision making what has to be obtained is not a compromise solution, but a preference structure similar to those of costumers with similar tastes. The problem is then not to simply combine the individual consumers preferences into an overall preference, but to combine the preferences privileging those of similar-minded people in the aggregation. As outlined in [6], these systems do not analyze the features or the descriptions of the products; rather they use consumers ratings to create a likability index for each product. This index is not global, but it is computed for an active user by means of the profiles of other users with similar interests. These techniques require the direct expression of users' preferences over the available products, generally expressed by means of grades of satisfaction. These systems are especially useful for the selection of commodity products (books, music, movies, etc.).

Another approach to give consumers a support in the activity of relevant products' identification is a descriptive one, based on an explicit description of the main features of the available products. The systems based on this approach are aimed at helping users to select the most interesting products among the available ones on the basis of an analysis of their characteristics. These systems are based on the so-called constraint satisfaction methods [6]. The techniques of *constraint-based filtering* use features of products to the aim of determining their relevance to consumers. If a customer is allowed to express constraints on the features of the products, the problem of product selection can be solved as a MCDM problem. In this case the attributes directly correspond to product features, and the constraints are expressed on these features.

Content-based approaches to product recommendations are based on an explicit customer's description of the preferred products, which has to be compared with the stored descriptions of the products. The usual way of specifying the customer's preferences is through the specification of keywords, like in Information Retrieval [6].

In this paper an approach which is between the content-based and the constraint-based approach is proposed. The aim of our method is to first cluster the set of products into a finite number of categories, based on the price feature; second this method provides an explicit description of the main features of the products, finalized at explicitly capturing a set of main constraints satisfied by a majority of the products belonging to a same cluster. In other words the proposed method realizes a linguistic categorization of the products: the products are first classified by a means of a clustering algorithm; each cluster is then described by means of a set of linguistic statements which qualify some characteristics (features) of the products. The aim of a qualitative classification of the products is to help the consumers to identify the products relevant to their needs. The proposed approach is described in sections 3 and 4.

3 The proposed method: the product space construction and clustering phases

In this section and in section 4 a novel approach for eliciting the main features characterizing categories of products is described. The aim of a qualitative classification of the products is to help the consumers to identify the products relevant to their needs.

The proposed approach consists of three distinct phases: in the preliminary phase, named the *Product Space Construction*, the description of the products by means of a set of their features is provided. This phase needs the support of a product-line expert, and provides as an output a synthetic description of the products, based on the values of the pre-defined set of features. In the second phase, the *Product Space Segmentation*, starting from the description of the product space a clustering algorithm is applied in order to partition the products into classes, on the basis of some specified feature. The price constitutes a primary selection criterion in purchasing activities; for this reason the proposed method applies a clustering based on this criterion; other criteria could however be considered to apply this a partition of the products into main "lines". The last phase (called the *Product Space Description*) is aimed at providing a qualitative summarization of the basic characteristics of the products in each class. This last phase is described in section 4.

3.1 Product space construction

Let us denote by $P = \{P_1, \ldots, P_N\}$ the set of the considered products which are described by means of a set of measurable characteristics or features $F = \{F_1, \ldots, F_M\}$. In particular, we consider characteristics such as price and technical or manufacturing features; if for example we consider cars as the products to be evaluated, some charateristics, which can be considered, are the cost, the top speed, and the gas consumption per kilometer. With each feature a reference domain is associated, which defines its possible values. On the basis of the values of the features of the products, a table is compiled a priori and stored on the site; this table gives a synthetic description of the products and it must be compiled by an expert in the product field. In Table 1 the rows refer to the products and the columns to the considered features. A value v_{ij} represents the value of the feature j for the product i.

Table 1. characteristic values of the products

	F_1	F_2	...	F_M
P_1	v_{11}	v_{12}		v_{1M}
P_2				
...				
P_N	v_{N1}	v_{N2}		v_{NM}

The construction of the product space is very important, as it produces a representation of the products which constitutes the knowledge interface between the consumer and the products themselves. The information collected in this phase is a necessary basis to automatically produce a summarization of some qualitative features of the products; the application of a clustering algorithm first, and the linguistic description of the main features shared by a soft majority of the elements belonging to a given cluster constitute a means to clarify and exploit a more general and synthetic knowledge of the products. This more general description can constitute a useful basis to direct the consumers in focusing their attention on the most relevant products.

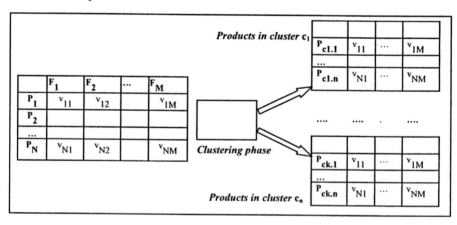

Fig.1: the result produced by the clustering phase on the product space description

Once the organization of the knowledge about the main features of the products has been completed, the subsequent phase of the proposed method is the *product space segmentation phase*; in order to facilitate the identification of the products satisfying the customer constraints, we propose to partition the products into clusters on the basis of some "leading" criterion. The basic idea is that a possible strategy aimed at the identification of the relevant products to customer's needs is to reduce the space of the products by directing the selection in a subset satisfying certain preliminary constraints. We think that a primary criterion which in most shopping activities guides the choices of a consumer is the price: generally a consumer has an idea of the threshold which he/she can't exceed. In section 3.2 we synthetically present the hierarchical clustering algorithm that we suggest to apply in order to partition the set of products into clusters. The output of the clustering phase is the partition of the initial table of the products features into n tables, in which n is the number of the generated clusters, as shown in Figure 1.

3.2 The clustering strategy

In this section we describe a method for clustering called Hierarchical Ag-Glomerative (HAG) clustering [13]. We shall use this method for segmenting the objects in a product line into different categories based upon price. This clustering method provides users with the capability of controlling the clustering process in way that allows for the easy inclusion of contextual knowledge, the satisfaction of various types of criteria by the cluster formation process and the possibility for the inclusion of human like intelligence to guide the clustering process [14].

Using this method clusters are built up by combining, based upon their proximity, existing clusters. The following algorithm summarizes the basic steps in the HAG clustering approach:
1. Select an initial set of clusters
2. Find the nearest pair of clusters and merge them
3. Repeat Step 2 until the stopping rule is satisfied

In each of these steps there exists a degree of freedom which allows an adaptation of the procedure to include various aspects of intelligence.

We start with the first step, here we must decide upon the initialization of the clusters. The most common way of initialization is to assign each object to a distinct cluster. However, there exists another way of initializing the clustering process called *directing* or *seeding*. In this process rather than making each of the initial clusters correspond to singletons we collect some of the objects into clusters. Using this approach we are providing some direction or guidance as to how the clusters will be formed. This is particular useful for imposing a requirement that certain objects are considered in the same cluster. In the second step of the algorithm, it is the selection of the method used to evaluate the proximity of clusters that provides us with our latitude in implementing the HAG clustering. In the third step, the stopping rule, we have not specified the process to be used. We can make some preliminary comments on how to do this. The stopping rule has to depend upon information available in the clustering process. One piece of information that we can use is the number of clusters. Thus we can specify some value N^* and indicate that when we have N^* clusters we stop, this is often used. Another piece of information that can be used is information about the number of elements in the clusters, the cluster cardinalities. For example, we can decide to stop if the cardinality of the biggest cluster is equal or greater than some specified value. Another method that is often used to determine the stopping rule is to stop the clustering process when the two nearest clusters are beyond a certain specified value. Combinations of these individual stopping criteria can be used *if the number of clusters is greater than N_1 and the minimal cluster distance is greater than D_1 then stop.*

In the algorithm used to implement the HAG clustering the basic process of cluster formation is the aggregation procedure, step two, in which we combine the two nearest clusters to form a new cluster. Fundamental to this process is the measurement of the proximity of two clusters. Even more basic than the measurement of the proximity of clusters is measurement of the proximity between individual observations. The measurement of proximity between object points is strongly problem dependent. Letting $X = \{x_1, x_2,..., x_n\}$ be our object set to be clustered, we denote the proximity of two objects x_i and x_j as $Prox(x_i, x_j)$. This measure can be of two types, "a distance type" or "a similarity type". By distance type measure, which we shall denote $Dist(x_i, x_j)$, we mean a proximity measure such that the smaller its value the nearer the objects. By a similarity type, which we shall denote $Sim(x_i, x_j)$, we mean a proximity measure such that the larger its value the nearer the objects. Some properties of the proximity measures are listed in [20].

In our particular problem, in which we are focusing on the clustering of objects with respect to their price, a natural measure of distance exists, it is the difference in price between two objects, $Dist(x_i, x_j) = |Price(x_i) - Price(x_2)|$. Thus for our purposes we shall most often view things from the perspective of a distance type measure, however we note a simple inversion operation allows us to convert a distance measure into a similarity measure. With this perspective in mind we shall assume that we have available an $n \times n$ matrix M such that m_{ij} is the distance between object x_i and object x_j, $m_{ij} = Dist(x_i, x_j)$. It is assumed, as required, that $m_{jj} = 0$ and $m_{ij} = m_{ji}$.

Extending the preceding notation if A and B are two clusters we shall let, $Dist(A, B)$ indicate the distance between these two clusters. We shall shortly turn to the issue of defining this measure of inter-distance. Using this concept of distance between clusters the basic step in the HAG clustering can now be formally described. Assume that at some point in the clustering process we have q clusters, C_k for $k = 1$ to q. We then select the clusters C_{i^*} and C_{j^*}, $i^* \neq j^*$, such that $Dist[C_{i^*}, C_{j^*}] = Min_{ij}[Dist(C_i, C_j)]$ to merge together. We shall call C_{i^*} and C_{j^*} the winning clusters.

We now turn to the issue of measuring the distance between two clusters. In the following we shall let A and B be two clusters. In particular A and B are two disjoint subsets of X, $A \cap B = \emptyset$. We shall denote $A = \{y_1,..., y_{na}\}$ and $B = \{z_1,..., z_{nb}\}$, here n_a and n_b are respectively the cardinality of A and B. As indicated we denote $Dist[A, B]$ as the proximity of the clusters A and B, thus the smaller $Dist[A, B]$ the closer the clusters. Since we have assumed the availability of the matrix M which contains the distance between any two elements in X we have available the distance between any element in A and any element in B, $Dist(y_i, z_j)$. In the following we let d_{ij} indicate the distance between y_i and z_j. We consider the question of obtaining $Dist(A, B)$ from the d_{ij}.

Two methods which have been often used for calculating the distance between two clusters are the nearest and furthest neighbor rules. The nearest neighbor rule defines the inter-cluster distance as the distance between the elements in each of the two clusters that are nearest,

$$Dist[A, B] = Min_{y_i \in A \ \& \ z_j \in B}[d_{ij}].$$

The furthest neighbor rule defines the inter-cluster distance as the distance between the elements in each of the two clusters that are furthest,

$$Dist[A, B] = Max_{y_i \in A \ \& \ z_j \in B}[d_{ij}].$$

We shall for simplicity denote these as Min[A, B] and Max[A, B] respectively.

Whichever of these methods we are using we are looking for the two clusters that are closest. In this respect some useful general observations can be made about these two approaches. Using the Min method we see that if A and A' are two clusters such that $A \subset A'$ then for any B

$$Min[A, B] \geq Min[A', B]$$

Hence when using the Min cluster distance measure the **more** elements in a cluster the more likely it is to be "nearer" to a neighboring cluster. This implies that this nearest neighbor clustering type approach tends to favor the formation of **big** clusters, at any point in the cluster merging process there is a propensity for the two largest clusters to unite. In this method there is an inclination for a few big clusters to form while the remaining elements will be in isolated small clusters. This is a kind of depth first clustering

Under the Max method if A and A' are two clusters such that $A \subset A'$ then for any other cluster B

$$Max[A, B] \leq Max[A', B]$$

The implications here is that the fewer the elements in the cluster the more likely it is to be "near" to a neighboring cluster. This method can be seen as a kind of "breath first" type clustering. Here there is a tendency to form clusters of uniform size and then joining these.

The two methods, the nearest and furthest neighbor method, can be seen as providing two extreme approaches to cluster aggregation. There exists another method, also commonly used. In this method we calculate the distance between two clusters by taking the average of distances between the elements in the two clusters: $Dist[A, B] = \dfrac{1}{n_a}\dfrac{1}{n_b}\sum_{i \in A}\sum_{j \in B} d_{ij}$. When using this method the number of elements in a cluster has no effect on the determination of Dist[A, B].

What we have noted in the preceding is the choice of inter-cluster distance measure clearly effects the process of cluster formation. A useful generalization of the preceding can be obtained. Let A and B be two clusters and let $R = \langle r_1 \ r_2,$ ...., $r_n \rangle$ be the collection of all distances between an element in A and an element in B, that is $r_k = Dist(y_i, z_j)$ for some $y_i \in A$ and $z_j \in B$. Thus $|A| = n_a$ and $|B| = n_b$

then $n = n_a \, n_b$. Then $G_\alpha(r_1, \ldots, r_n) = \left(\dfrac{1}{n} \displaystyle\sum_{k=1}^{n} r_k^\alpha \right)^{\frac{1}{\alpha}}$ provides a whole family of

cluster distance measures, Dist(A, B) where $\alpha \in [-\infty, \infty]$ is a parameter which determines a specific member of this family. Special members of this family are Min for $\alpha = -\infty$. Average for $\alpha = 1$ and Max for $\alpha = \infty$. As indicated in [20] choice of α can be seen as the selection of a parameter determining the process of cluster formation.

In [14,20] Yager suggested using the consequences of selecting different α's to control the cluster formation process by dynamically changing α, based upon information available about the process taking place and context dependent criteria desired to be satisfied in the clustering. In particular with this type of approach it is possible to introduce a mechanism to guide the formulation of clusters which can begin to manifest the kind intelligence exhibited by a human being.

4 The proposed method: the product subspaces descriptions through linguistic summaries

As outlined in section 3, the result of the clustering algorithm on the product space description is its decomposition into a finite set of subspaces, one for each cluster (as illustrated in Figure 1 above). The third phase of the consumer decision support method proposed in this paper, is to associate with each cluster some information summarizing the main features shared by the products belonging to that class. The aim of our method is to automatically construct for each cluster a set of linguistic assessments, which summarize the main "features" shared by a fuzzy majority of the products belonging to a class. An example of a such a linguistic assessment is "*Most* of the cars in class A have a *high* speed" in which the considered products are *cars* and the evaluated feature is the *speed*. The formal framework in which the third phase is modelled is constituted by fuzzy set theory; in particular linguistic summaries.

In section 4.1 we shortly introduce linguistic summaries, and in section 4.2 a method for linguistically qualifying the features of a fuzzy majority of the products contained in a given cluster is defined.

4.1 Fuzzy Set Theory and Linguistic Summaries

Fuzzy set theory has been introduced for modeling the graduality of the concept of membership of elements to a class, to the main aim of representing linguistic categories of the language (vague concepts) [22,23]. Fuzzy sets allow to represent various types of information (precise, imprecise – *between 20 and 30*, linguistic – *tall, small*, …), and they can be employed with different semantics (similarity, preference and uncertainty [4]). In [5] the notion of fuzzy information engineering has been introduced, as a set of methods useful to clarify, retrieve and exploit information. The approach which we propose in section 2.2 to define linguistic characterizations of the products in an E-commerce site, is based on some methods typical of fuzzy information engineering, and it has the aim of clarifying and exploiting some information concerning the products collected in a web-shopping site.

One of the possible semantics of fuzzy set is the one of flexible constraint, which is satisfied by a given element of the domain of discourse at a degree corresponding to its membership value [4]. For example a person having an height of m. 1.80 satisfies the constraint tall referred to the height of a person at a degree 0.8. The concept of flexible constraint plays a central role in defining the so-called linguistic summaries as we will see further in this section. Another concept which is central to linguistic summaries is the concept of linguistic quantifier: the concept of linguistic quantifier has been introduced to generalize the existential and universal quantifiers of classical logic [19,21]. They can be either crisp (such as for example *all, at least 1, at least k, half*) or fuzzy quantifiers (such as for example *most, several, some, approximately k*). In [15] linguistic quantifiers have been formalized by Ordered Weighted Averaging operators (OWA) to the aim of aggregating the satisfaction values of flexible constraints.

The notion of linguistic summary has been introduced as a means to synthesize some aspects of the information contained in a database where a large collection of data is stored [17,18]. A linguistic summary is a statement about a property of a set of data of the type "Q objects in D are S", in which Q is called the quantity in agreement and S is called the summarizer (it is a flexible constraint); we assume that Q is expressed by a monotone non-decreasing linguistic quantifier such as *most* or *at least n%* and S is expressed by a fuzzy set. An example of such a statement is "*most* of the employees in D are *young*". With a linguistic summary a degree of satisfaction can be associated, which expresses the coherence of the statement with respect to the information contained in the database (it reflects how much the information contained in the database satisfies the constraint expressed in the statement).

4.2 The linguistic summarization of product categories

The third phase of the proposed decision support method has the aim of qualifying the features of the products in the various clusters by associating with each cluster a set of linguistic statements, which describe some qualitative characteristics shared by a fuzzy majority of the products in the considered cluster.

The concept of fuzzy majority has been formalized in group decision making, in order to define the degree of consensus among a soft majority of the experts when an unanimous consensus cannot be reached [1,3,8,9,16]. The proposed structure of the statements to be associated with a cluster is the following: "Q products in this cluster have L feature-x", in which Q is a linguistic quantifier such as *most, at least 50%*, etc., feature-x is a feature of the considered products (for example the speed in the case of cars), L is a linguistic label qualifying the values of that feature (it expresses a flexible constraint satisfied by the precise values of the features of the products) [2]. The linguistic quantifiers allow the expression of a more flexible characterization of the contents of a cluster; a statement containing a linguistic quantifier reflects a property shared by the soft majority which it expresses. In order to automatically associate a set of qualitative statements to the products in a cluster the notion of linguistic summary is applied. In order to do this a set G of linguistic labels is identified for each considered feature, $G = \{G_0, G_1, ..., G_L\}$. The labels G_i identify flexible constraints on the feature values of the products and give a vague characterization of the products' feature. An example of such a set for the feature *top speed* of a car could be: {*very low, low, medium, high, very high*}. An ordinal scale is defined on the linguistic labels; this means that G_{l+1} is a better evaluation than G_l, $\forall l \in [0, ..., L]$. The membership functions of the labels have to be distinctly defined for each feature, as they refer to distinct valued domains (for example the domain of car's speed can be defined as the numeric interval [0, 220] (km/h)). The linguistic labels identify flexible constraints that the feature values stored in Table 1 satisfy at a certain degree, which is the membership degree of the precise feature value in the fuzzy set associated with the linguistic evaluation label. For example let consider a given car, say a car of model Fiat Tipo, and let also consider the feature *maximum speed* of cars, which can be specified by a number in the interval [180,280] km/h and which is linguistically characterized by means of a value of the following set $G_{speed} = \{low, medium, high\}$. If the maximum speed of our Fiat Tipo car is, say, 220 km/hour, the considered car satisfies the flexible constraints imposed by the linguistic values of speed at a distinct extent, depending on the definition of their membership functions. These satisfaction values are in fact the membership degrees of the numeric value 220 to the fuzzy sets associated with the linguistic labels. In figure 2 an example of such functions is shown.

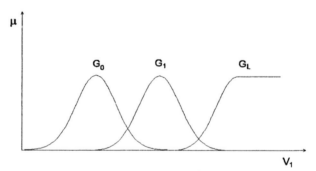

Fig. 2: examples of membership functions

In order to identify a linguistic value characterizing the top speed of the Fiat Tipo car, we can select the one with the highest membership degree in correspondence of the value 220 (the constraint which is better satisfied); in the example above let us assume that this maximum membership degree is the one to the fuzzy set associated with the concept *high*. In order to linguistically characterize the value of each feature by a majority of the elements contained in a cluster, a set of linguistic summaries is defined, one for each linguistic value referred to the considered feature; the linguistic value which summarizes the feature value by a soft majority of the elements in a cluster is the one contained in the linguistic summary with the highest satisfaction degree. More formally, for a given cluster and a given feature the following procedure is adopted to define the linguistic characterization shared by a majority of the product contained in the cluster:

- for each feature a set of qualifying linguistic labels is defined, together with their corresponding fuzzy sets;
- for each feature a set of L general statements (summaries) is defined of the type "Q products in cluster Cl have a Y *feature*", in which Y varies in G and L is the number of elements in the set G;
- in order to select the linguistic characterizion of each feature in a given cluster Cl the procedure described in the following is applied. It consists in verifying the degree of satisfaction of the L linguistic summaries, and selecting the one with the maximum satisfaction level.

This procedure is applied for each of the M considered features, as it is sketched in the following table.

Table 2. Linguistic summaries describing the products' features for a given cluster

For feature F_1
Q products have a G_0 *feature* F_1
Q products have a G_1 *feature* F_1
....
Q products have a G_L *feature* F_1
....
For feature F_M
Q products have a G_0 *feature* F_1
Q products have a G_1 *feature* F_1
....
Q products have a G_L *feature* F_1

Example: in the case of the feature speed of a cluster Cl of cars the following summaries have to be verified:

> *most* of the cars in cluster Cl have a *low* speed
> ....
> *most* of the cars in cluster Cl have a *medium* speed
> *most* of the cars in cluster Cl have a *high* speed

For each product in the cluster and each feature the membership degrees in the fuzzy sets associated with the linguistic labels qualifying the feature are computed; in this way a numeric degree for each product, each feature and each evaluation label is obtained. In Table 3 a scheme is presented which is referred to a considered cluster Cl containing N_{Cl} products. For a given feature F_i there are L rows, where L is the number of the linguistic labels specifying the flexible constraints on that feature; each column is referred to a product of the considered cluster, and the value a_{ijk} refers to the satisfaction degree of constraint imposed by G_k on the feature F_i of the product P_j. In order to compute an overall satisfaction degree of the flexible constraint G_k by a majority Q of the products, the values in the row corresponding to the value G_k for the feature F_i have to be aggregated.

Table 3. Computation of the satisfaction degrees α_{ij} of the flexible constraints G_j by a majority Q of the products for the feature F_i

		P_1	P_2	...	P_{NCl}	
F_1	G_0	a_{110}	a_{1200}	$a_{1.0}$	a_{1N0}	→ α_{10}
	G_1					→ α_{11}
	...					...
	G_L	a_{11L}	a_{12L}	$a_{1..L}$	a_{1NL}	→ α_{1L}
...	G_0					
	G_1					
	...					
	G_L					
F_M	G_0					→ α_{M0}
	G_1	a_{M11}	a_{M21}	$a_{M.1}$	a_{MNiN1}	→ α_{M1}
	...					...
	G_L	a_{MIL}	a_{M2L}	$a_{M.L}$	a_{MNL}	→ α_{ML}

This aggregation produces the degree of satisfaction of a considered linguistic summary. The aggregation is performed by means of the aggregation operator associated with the linguistic quantifier Q; to this aim in [15] Ordered Weighed Averaging Operators have been employed. Once the aggregated values have been obtained, for each feature the linguistic summary with the highest satisfaction degree is selected as the proposition summarizing the feature-based qualification.

In Table 4 the procedure for the selecting the linguistic summary which better characterizes a given feature F_i is sketched.

Table 4. selection of a linguistic summary on the basis of the satisfaction degrees

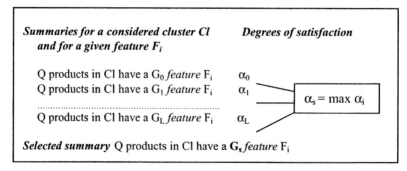

5   Conclusions

In this paper a method to define a linguistic categorization of the products in a web-shopping site has been described. This method is essentially based on clustering a product line into a number of categories, typically defined by price, along with defining a linguistic description of the relevant features of the products associated with each category. The aim of this method is to allow consumers to easily get a good understanding of the issues involved choosing a product in a given category. The definition of linguistic descriptions make use of some concepts defined within fuzzy set theory.

References

1. Bordogna G., Fedrizzi M., and Pasi G., "A linguistic modeling of consensus in Group Decision Making based on OWA operators", IEEE Trans. on System Man and Cybernetics, 27 (1), 1997.
2. Cescotti S., Pasi G., "*A model of group decision making based on linguistic summaries*", *proc.* Information Processing and Management of Uncertainty in Knowledge Based Systems IPMU *2000, Madrid, July 2000.*
3. Consensus under fuzziness, J.Kacprzyk, H. Nurmi, M. Fedrizzi eds, 1997.
4. Dubois D., and Prade H., The three semantics of fuzzy sets, Fuzzy Sets and Systems, 90, 141-150, 1997.
5. Fuzzy Information Engineering: a Guided Tour of Applications, D. Dubois, H. Prade and R.R. Yager eds., John Wiley and Sons, 1996.
6. Guttman R.H., Moukas A.G., Maes P., Agent-mediated Electronic Commerce: a Survey. *Knowledge Engineering Review*, 13(2), pages 147-159, 1998.
7. Häubl G., Trifts V., Consumer Decision Making in Online Shopping Environments: The Effects of Interactive Decision Aids. *Marketing Science*, 19(1), pages 4-21, 2000.
8. Kacprzyk J. and Roubens M., (eds) "Non conventional Preference Relations in Decision Making", Springer-Verlag, 1988.
9. Kacprzyk J., Fedrizzi M. and Nurmi H., "Group Decision Making and Consensus under Fuzzy Preferences and Fuzzy Majority", Fuzzy Sets and Systems, 49, 21-31, 1992.
10. Pennock D.M., Horvitz E., Lawrence S., Lee Giles C., Collaborative filtering by personality diagnosis: a hybrid memory and model-based approach. In *Proceedings of the 16th Conference on Uncertainty in Artificial Intelligence*, UAI 2000, Stanford, CA, June 2000.
11. Perny P. and Zucker J.D., Collaborative Filtering Methods based on Fuzzy Preference Relations. In *Proceedings of the conference EUROFUSE-SIC'99*, Budapest, 1999.
12. Reisnick P. and Varian H.R., Recommender Systems. Special issue of *Communications of the ACM*, 40(3), 1997.
13. Sokal, R. R. and Sneath, P. H. A., Principles of Numerical Taxonomy, W. H. Freeman: San Francisco, 1963.

14. Yager, R. R., "On the control of hierarchical clustering," Proceedings Eight IPMU International Conference on Information Processing and Management of Uncertainty in Knowledge-Based Systems, Madrid, 1335-1341, 2000.

15. Yager R. R., "On Ordered Weighted Averaging aggregation Operators in Multi Criteria Decision Making", *IEEE Trans. on Systems, Man and Cybernetics* 18(1), 183-190, 1988.

16. Yager R.R., Non Numeric Multi Criteria Multi-Person Decision Making, Group Decision and Negotiation, 2, 81-93, 1993.

17. Yager R.R., "A New Approach to the Summarization of Data", Information Sciences, 28, 69-86, 1982.

18. Yager R.R., "On Linguistic Summary of Data", in Knowledge Discovery in Database, edited by Piatetsky – Shapiro G., Frawley W. J., 347-363, 1991.

19. Yager R.R., Interpreting Linguistically Quantified Propositions, in International Journal of Intelligent Systems, Vol. 9, 541-569, 1994.

20. Yager R.R., Intelligent Control of the Hierarchical Agglomerative Clustering Process", IEEE Transactions on Systems, Man and Cybernetics: Part B 30, 835-845, 2000.

21. Zadeh L.A., A computational Approach to Fuzzy Quantifiers in Natural Languages, *Computing and Mathematics with Applications.* 9, 149-184, 1983.

22. Zadeh L.A., The Concept of a Linguistic Variable and its Application to Approximate Reasoning parts I and II, in Information Sciences, n.8, 199-249, 301-357, 1975.

23. Zadeh L.A., "Fuzzy Logic = Computing with Words", IEEE Transactions on Fuzzy Systems, 4, 103-111, 1996.

Logo Recognition and Detection with Geostatistical, Stochastic, and Soft-Computing Models

Tuan D. Pham

Command and Control Division, Defence Science and Technology Organisation,
PO Box 1500, Edinburgh, SA 5111, Australia.
tuan.pham@dsto.defence.gov.au

1 Introduction

To meet an increasing demand for the automatic processing of office documents, logo detection and recognition in document images play an important role in this area. As an instance, given a large volume of faxed documents which need to be classified into different sources of topics. Therefore if the topic of the incoming information can be automatically identified by the logo the electronic document bears, then the information analyst can be effectively aided to make a decision whether or not to analyse the context of the document.

Logo recognition and logo retrieval share the same mathematical methods and algorithms but have different criteria that can be described as follows.

Let $\mathbf{I} = \{I(x,y) : x = 1, 2, \ldots, M; y = 1, 2, \ldots, N\}$ be a two-dimensional image, and g is a function that maps the image space $I(x,y)$ onto the feature vector $\mathbf{s} = \{s_1, s_2, \ldots, s_T\}$:

$$g : \mathbf{I} \to \mathbf{s}.$$

The measure of similarity between two images $\mathbf{I}_1$ and $\mathbf{I}_2$ can be expressed in terms of their feature vectors using a function d such as a distance measure or a scoring model.

If $\mathbf{Q}$ is a query image, the task of image retrieval is to obtain a collection of images $\mathcal{C}$ in the image database $\mathcal{L}$ that are similar to $\mathbf{Q}$ based on the following criterion:

$$d[g(\mathbf{Q}), g(\mathbf{I})] \le \delta, \ \mathbf{I} \in \mathcal{C}$$

where δ is a prescribed threshold.

For the task of image recognition, an unknown image $\mathbf{U} \in \mathcal{L}$ identified as image $\mathbf{I}^* \in \mathcal{L}$ can be determined by the following criterion:

$$I^* = \arg\max_i d[g(\mathbf{U}), g(\mathbf{I}_i)], \ \mathbf{I}_i \in \mathcal{L}$$

where the maximum value of the similarity measure indicates the maximum matching degree between the unknown image and the image in the database.

In the context of retrieval or recognition of trademark images, a number of methods have been proposed for such tasks. Some of these investigations include the measure of similarity of trademark shapes by string-matching techniques [7] that chaincodes the object contours and applies string distances to the coded contours in order to measure the shape similarity of 2-D objects. A multi-level or hierarchical approach to logo recognition was developed [9, 10] which uses text and contour features to prune the database and similarity invariants to obtain a more refined match. Another hierarchical shaped-based method [20] for the retrieval of trademark images uses edge directions and invariant moments as the extracted features of trademark images for matching and integration, then deformable templates are further applied to enhance the shortcomings of the feature-level matching. As another approach, a shape-feature based method using closed contours was also proposed for trademark recognition [31]. A pseudo hidden Markov model (HMM) was recently developed for retrieving deformed trademarks [4]. This method incorporates color components and individual silhouettes as shape features of trademarks into a 2-D pseudo HMM. An approach using relevance feedback for content-based similarity retrieval of trademarks was proposed [6]. This method is based on the hypothesis that low-level image features used to index trademark images can be correlated with the image contents by applying a relevance feedback to evaluate the feature distributions and then the similarity measure and query are updated accordingly to enhance the representation of the user's needs.

Methods for the recognition and retrieval of other image shapes include polygon approximation and neural network [28]; neural-net classifiers [37, 24]; shape matching and similarity ranking in eigenshape space for image retrieval [14]; higher-order spectra invariants [3, 23, 40]; segmentation of color-image regions for content-based image retrieval [13]; survey of shape analysis techniques [26] including boundary transform techniques, boundary space domain techniques, global scalar transform techniques, and global space techniques. Furthermore, other investigations on extraction of invariant features for pattern recognition and image retrieval can also be found in references [5, 2, 12, 16, 17, 22, 36] and also in [39, 43, 44, 47, 49, 50].

Feature extraction by the methods of moments are among the most popular choices for image recognition and retrieval [5]. However, in the domain of logo recognition, the invariant moments and other shape features are not very effective in retrieving rotated images, and there seems that no single feature can effectively recognize trademark images subjected to noise and several geometrical distortions such as scaling, translation, and rotation [20]. This paper presents a new approach for extracting features of logo images based on the spatial second-order moment of geostatistics known as the variogram whose experimental measures can be effectively learned by a neural network for recognizing logos under geometrical changes and noise.

As a component of a fully automated logo recognition system, the detection of logos in electronic documents is carried out first in order to determine the existence of a logo that will then be classified to best match the logo in the database. Computer methods for logo detection have been rarely reported in the literature of document imaging. In the published literature of logo detection we have found but a single work by Seiden et al [38] who developed a detection system by segmenting the document image into smaller images which consist of several document segments such as small and large texts, picture, and logo. The segmentation is based on a top-down, hierarchical X-Y tree structure [29]. Sixteen statistical features are extracted from these segments and a set of rules is then derived using the ID3 algorithm [32] to classify if an unknown region is likely to contain a logo or not. This detection algorithm is also independent of any specific logo database as well as the location of the logo.

The following sections present new methods for extracting features of intensity logo images, three logo classifiers, and a logo detector. The public-domain logo database of the University of Maryland is used to illustrate the performance of the proposed methods.

2 Feature Extraction

The theory of regionalized variables [27] states that when a variable is distributed in space, it is said to be regionalized. A regionalized variable is a function which takes a value at point p of coordinates (p_x, p_y, p_z) in three-dimensional space and consists of two conflicting characteristics in both local erratic and average spatially structured behaviors. The first behavior yields to the concept of a random variable; whereas the second behavior requires a functional representation [21]. In other words, at a local point p_1, $F(p_1)$ is a random variable; and for each pair of points separating from a spatial distance h, the corresponding random variables $F(p_1)$ and $F(p_1 + h)$ are not independent but related by the spatial structure of the initial regionalized variable.

By the hypothesis of stationarity [21, 15], if the distribution of $F(p)$ has a mathematical expectation for the first-order moment, then this expectation is a function of p and expressed by

$$E\{F(p)\} = \mu(p) \tag{1}$$

The three second-order moments considered in geostatistics are as follows.

1. The variance of the random variable $F(p)$:

$$Var\{F(p)\} = E\{[F(p) - \mu(p)]^2\} \tag{2}$$

2. The covariance:

$$C(p_1, p_2) = E\{[F(p_1) - \mu(p_1)][F(p_2) - \mu(p_2)]\} \tag{3}$$

3. The variogram function:

$$2\gamma(p_1, p_2) = Var\{F(p_1) - F(p_2)\} \tag{4}$$

which is defined as the variance of the increment $[F(p_1) - F(p_2)]$. The function $\gamma(p_1, p_2)$ is therefore called the semi-variogram.

The random function considered in geostatistics is imposed with the four degrees of stationarity known as strict stationarity, second-order stationarity, the intrinsic hypothesis, and quasi-stationarity. Strict stationarity requires the spatial law of a random function, that is defined as all distribution functions for all possible points in a region of interest, is invariant under translation. In mathematical terms, any two k-component vectorial random variables $\{F(p_1), F(p_2), \ldots, F(p_k)\}$ and $\{F(p_1+h), F(p_2+h), \ldots, F(p_k+h)\}$ are identical in the spatial law whatever the translation h. The second-order stationarity possesses the following properties:

1. The expectation $E\{F(p)\} = \mu(p)$ does not depend on p, and is invariant across the region of interest.
2. The covariance depends only on separation distance h:

$$C(h) = E\{F(p+h)F(p)\} - \mu^2, \forall p \tag{5}$$

where h is a vector of coordinates in one to three dimensional space.
If the covariance $C(h)$ is stationary, the variance and the variogram are also stationary:

$$\begin{aligned} Var\{F(p)\} &= E\{[F(p) - \mu]^2\} \\ &= C(0), \forall p \end{aligned} \tag{6}$$

$$\begin{aligned} \gamma(h) &= \frac{1}{2}E\{[F(p+h) - F(p)]^2\} \\ &= C(0) - C(h). \end{aligned} \tag{7}$$

The intrinsic hypothesis of a random function $F(p)$ requires that the expected values of the first moment and the variogram are invariant with respect to p. That is the increment $[F(p+h) - F(p)]$ has a finite variance which does not depend on p:

$$\begin{aligned} Var\{F(p+h) - F(p)\} &= E\{[F(p+h) - F(p)]^2\} \\ &= 2\gamma(h), \forall p \end{aligned} \tag{8}$$

The quasi-stationarity is defined as a local stationarity when the maximum distance $|h| = \sqrt{h_x^2 + h_y^2 + h_z^2} \leq b$. This is a case where two random variables $F(p_k)$ and $F(p_k + h)$ cannot be considered as coming from the same homogeneous region if $|h| > b$.

Let $f(p) \in \Re$ be a realization of the random variable or function $F(p)$, and $f(p + h)$ be another realization of $F(p)$ separated by the vector h. Based on (4), the variability between $f(p)$ and $f(p+h)$ is characterized by the variogram function:

$$2\gamma(p, h) = E\{[F(p) - F(p + h)]^2\} \tag{9}$$

which is a function of both point p and vector h, and its estimation requires several realizations of the pair of random variables $[F(p), F(p + h)]$.

In many applications, only one realization $[f(p), f(p+h)]$ can be available, that is the actual measure of the values at point p and $p+h$. However, based on the intrinsic hypothesis, the variogram $2\gamma(p, h)$ is reduced to the dependency of only the modulus and direction of h and does not depend on the location p. The semi-variogram $\gamma(h)$ is then constructed using the actual data as follows.

$$\gamma(h) = \gamma(r, \theta) = \frac{1}{2N(h)} \sum_{i=1}^{N(h)} [f(p_i) - f(p_i + h)]^2 \tag{10}$$

where $r = |h|$ and θ are the modulus and orientation of h respectively. $N(h)$ is the number of experimental pairs $[f(p_i) - f(p_i + h)]$ of data separated by h. Based on this notion, this function $\gamma(h)$ is said to be the experimental semi-variogram.

There are several mathematical versions for modeling the theoretical semi-variograms [19] that allow the computation of a variogram value for any possible distance h. The three most commonly used theoretical semi-variogram models are given as follows [19, 11].

1. The spherical model:

$$\gamma(h) = \begin{cases} 1.5\frac{h}{a} - 0.5(\frac{h}{a})^3 & : \quad h \leq a \\ 1 & : \quad otherwise \end{cases} \tag{11}$$

   where a is called the range at which $\gamma(h)$ reaches a plateau. The range a can be estimated from the experimental semi-variogram.

2. The exponential model:

$$\gamma(h) = 1 - \exp\left(-\frac{3h}{a}\right) \tag{12}$$

3. The Gaussian model:

$$\gamma(h) = 1 - \exp\left(-\frac{3h^2}{a^2}\right) \tag{13}$$

However, in order to fit an experimental variogram into any of the above theoretical variogram models, manual fitting and subjective judgment are usually required to ensure the validity of the model. To overcome this problem that is necessary for the task of automatic recognition, we have chosen to use neural networks to learn the variogram functions from their experimental data.

Trademark images are characterized mostly by their spatial structures, the use of the variogram therefore will be helpful to capture the spatial relationships of content-based image data. An approximation of the experimental semi-variogram for 2-D images, that works well in practice, is given by [35, 42]

$$\gamma(h) = \frac{1}{2} \left(\frac{1}{N_r N_c} \sum_{x=1}^{N_c} \sum_{y=1}^{N_r} [I(x+h,y) - I(x,y)]^2 \right.$$
$$\left. + \frac{1}{N_r N_c} \sum_{x=1}^{N_c} \sum_{y=1}^{N_r} [I(x,y+h) - I(x,y)]^2 \right) \tag{14}$$

where $I(\cdot)$ is the pixel value, N_r and N_c are the number of rows and columns of the image respective, and h is the distance expressed in pixels.

3 Neural-Network based Classifier

Neural networks have been well known for their capabilities for function approximation and applied herein for approximating the variogram functions from discrete values of the experimental variograms. The multi-layer feedforward neural network receives the values of experimental variograms $\gamma(h)$ defined in (14) as inputs where the number of distances h are determined when the range a is reached. The output layer consists of the number of reference trademarks and responses with an integer 1 in the position of the trained trademark and 0 for the other positions.

The logistic sigmoid transfer function is selected because it interprets the network outputs as posterior probabilities that can produce very powerful results in terms of discriminancy [1]. The logistic sigmoid transfer function is given by

$$s(\beta) = \frac{1}{1 + \exp(-\beta)} \tag{15}$$

which generates an S-shaped curve and maps the input interval $(-\infty, \infty)$ onto $[0, 1]$. If $|\beta|$ is small then $s(\beta)$ is approximated by a linear function.

4 Pseudo HMM based Classifier

The elements of a hidden Markov model $\lambda = (A, B, \pi)$ are defined as [33, 34]

- N: the number of hidden states.
- M: the number of observation symbols.
- $A = \{a_{ij}\}$: the state-transition probability distribution, in which

$$a_{ij} = P(q_{t+1} = j | q_t = i), \ 1 \leq i, j, \leq N$$

where q_t denotes the state at time t.
- $B = \{b_j(k)\}$: the obervation symbol distribution, in which $b_j(k)$ is the symbol distribution in state $j, j = 1, \ldots, N$. It can be expressed as

$$b_j(k) = P(o_t = v_k | q_t = j), \ 1 \leq k \leq M$$

where v_k denotes an individual symbol.
- $\pi = \{\pi_i\}$: the initial state distribution, which is expressed as

$$\pi_i = P(q_1 = i), \ 1 \leq i \leq N$$

Three basic problems of interest for HMMs to be useful in real-world applications are: Scoring a model against a given observation sequence; discovering the hidden states of the model, and optimizing the model parameters to best describe the occurance of the observations. In this study, we are interested in the scoring problem that can be expressed in terms of an HMM as follows.

Given the sequence of observation symbols $O = < o_1, o_2, \ldots, o_T >$, and the model $\lambda = (A, B, \pi)$, it is then to compute the probability of the observation sequence $P(O|\lambda)$.

The forward and backward algorithms were developed to solve such scoring problem. Given a hidden Markov model λ, the forward variable, denoted by $\alpha_t(i)$, is the probability of partial observation sequence $< o_1, o_2, \ldots, o_t >$ and state i at time t:

$$\alpha_t(i) = P(o_1 o_2 \ldots, q_t = 1 | \lambda) \tag{16}$$

The term $\alpha_t(i)$ can be solved by induction as

Initialization
$$\alpha_1(i) = \pi_i b_i(o_1), \ 1 \leq i \leq N \tag{17}$$

Induction
$$\alpha_{t+1}(j) = \left[\sum_{i=1}^{N} \alpha_t(i) a_{ij} \right] b_j(o_{t+1}), \tag{18}$$

$1 \leq t \leq T - 1; \ 1 \leq j \leq N$

Termination
$$P(O|\lambda) = \sum_{i=1}^{N} \alpha_T(i) \tag{19}$$

The backward procedure can also be constructed for calculating $P(O|\lambda)$. Given the model λ and state i at time t, the backward variable, denoted by $\beta_t(i)$, is defined as the probability of partial observation sequence from time $t+1$ to the end of the sequence. Once again using the induction method, the backward term $\beta_t(i)$ can be solved as

Initialization

$$\beta_T(i) = 1, \ 1 \leq i \leq N \tag{20}$$

Induction

$$\beta_t(i) = \sum_{j=1}^{N} a_{ij} b_j(o_{t+1}) \beta_{t+1}(j), \tag{21}$$

$t = T-1, T-1, \ldots, 1; \ 1 \leq i \leq N$
Termination

$$P(O|\lambda) = \sum_{i=1}^{N} \beta_1(i) \tag{22}$$

We extract the first M lags of the experimental semi-variograms and build each discrete HMM for each logo where the number of states N is equal to the number the semi-variogram lags, that is $N = M$.

Due to our prior knowledge of the semi-vriogram distribution, the initial state probability distribution is defined as

$$\pi_i = \begin{cases} 1 & : \quad i = 1 \\ 0 & : \quad \text{otherwise} \end{cases}$$

where $1 \leq i \leq N$.

Similarly the state transition distribution is given by

$$a_{ij} = \begin{cases} 1 & : \quad j = i+1 \\ 0 & : \quad \text{otherwise} \end{cases}$$

where $1 \leq i, j \leq N$.

The observation symbol probability distribution B is obtained using the LBG algorithm for vector quantization (VQ) [25], which has been widely used in speech and speaker recognition. For the VQ design, 8 codewords are built to generate the symbol sequences. However, due to the limitation of the logo database under study, in which there is only one sample for each original image, we therefore build different observation symbol distributions according to each original image, its geometrical orientations, and noisy versions and then select the model probability by

$$P(O|\lambda^\kappa) = \arg \max_{1 \leq v \leq V} [P(O|A^\kappa, B_v^\kappa, \pi^\kappa)] \tag{23}$$

where $\kappa = 1, \ldots, \mathcal{K}$, the total number of logos, and V is the number of observation distributions of λ^κ.

284

We can see from now that this is but building a pseudo HMM for each logo since the state distribution is deterministic as given in (4) rather than a probability distribution. This is due to the fact that we already know the spatial distribution of the semi-variogram function, that is $\gamma(1), \gamma(2), \ldots, \gamma(h = N)$, which therefore yields one such fixed state sequence

$$Q = < q_{t=1}, q_{t=2}, \ldots, q_{t=N} >$$

which leads to a single-path scoring without the use of the forward or backward algorithm. Furthermore, there is also no need for re-estimating λ since both π and A are deterministic.

Assuming statistical independence of the observations, the scoring can be simply obtained by calculating the likelihood

$$P(O|Q, \lambda) = \prod_{t=1}^{N} P(O|q_t, \lambda)$$

that is

$$P(O|Q, \lambda) = b_{q_1}(o_1) b_{q_2}(o_2) \ldots b_{q_N}(o_N) \tag{24}$$

Using the logarithm, we have

$$\log[P(O|Q, \lambda)] = \sum_{t=1}^{N} \log[b_{q_t}(o_t)] \tag{25}$$

in which $b_{q_t}(o_t)$ is set to be 0.000001 if $b_{q_t}(o_t) = 0$.

We then assign the logo to class κ^* according to the model that has the maximum log-likelihood as follows.

$$\kappa^* = \arg\max_{\kappa}\{\log[P(O|Q, \lambda^\kappa)]\} \tag{26}$$

5 Fuzzy-Set based Classifier

It can be seen in (24) or (25) that the use of vector quantization is not very effective in the calculation of the probability of sequential observations when the training data is insufficient. This is true for the problem of logo recognition since there are not many logo samples of the same class for training. In fact, these probabilities are inherently imprecise or fuzzy because they are not representative of a reasonable random outcome. In case an unforseen (noisy) pattern is encountered, some codewords of this pattern are easily generated differently in comparison with its trained pattern, which are caused by the hard partition of the k-means clustering. Therefore, fuzzy vector quantization (FVQ) [45] was introduced to handle such cases. In this problem, there is no need for the re-estimation of λ in the pseudo HMM. Thus, FVQ is not applied herein but we model each codeword as a fuzzy set that allows a semi-variogram value to belong to more than one codeword, and the classification is based on the modification of (24).

Let $(R^n, \mathcal{A}, P)$ be a probability space in which $\mathcal{A}$ is a σ-field of Borel sets in R^n and P is a probability measure over R^n. Then, a fuzzy event in R^n is a fuzzy set S in R^n whose membership function $\mu_S(x)$ is a Borel measure. Then, the probabilty of a fuzzy event S is defined by the Lebesgue-Stieltjes integral as [51]

$$\tilde{P}(S) = \int_{R^n} \mu_S(x)dP = E(\mu_S) \tag{27}$$

by which the probability of a fuzzy event is the expectation of its membership function.

Let $c_1 < c_2 \ldots < c_M$ be the codewords obtained from a vector quantization (we choose not to denote these codewords in conventional bold type because they have only one dimension), and also let $\omega_l = c_{k-1}$, $\omega_m = c_k$, and $\omega_r = c_{k+1}$, $1 < k < M$.

We use the linear and triangular functions to obtain the fuzzy membership grades for the observation symbols at times $t = 1$, $1 < t < N$, and $t = N$ with respect to the semi-variogram γ_t as follows.

$$\mu(\gamma_t, c_{k=1}) = \begin{cases} 1 & : \quad \gamma_t \leq \omega_m \\ \dfrac{\omega_r - \gamma_t}{\omega_r - \omega_m} & : \quad \omega_m \leq \gamma_t \leq \omega_r \end{cases} \tag{28}$$

$$\mu(\gamma_t, c_{1<k<M}) = \begin{cases} 0 & : \quad \gamma_t \leq \omega_l \\ \dfrac{\gamma_t - \omega_l}{\omega_m - \omega_l} & : \quad \omega_l \leq \gamma_t \leq \omega_m \\ \dfrac{\omega_r - \gamma_t}{\omega_r - \omega_m} & : \quad \omega_m \leq \gamma_t \leq \omega_r \\ 0 & : \quad \omega_r \leq \gamma_t \end{cases} \tag{29}$$

$$\mu(\gamma_t, c_{k=M}) = \begin{cases} \dfrac{\gamma_t - \omega_l}{\omega_m - \omega_l} & : \quad \omega_l \leq \gamma_t \leq \omega_m \\ 1 & : \quad \omega_m \leq \gamma_t \end{cases} \tag{30}$$

The fuzzy scoring, denoted by $\tilde{P}(O|Q, \lambda)$, is computed by taking into account the overlapping between adjacent codewords as follows.

$$\tilde{P}(O|Q, \lambda) = \prod_{t}^{N} \psi_t \tag{31}$$

where ψ_t is the probability of the fuzzy observation $b_{q_t}(o_t)$, and defined according to (27) as follows.

$$\psi_t = \sum_{j \in J} b_{q_t}(o_j)\, \mu(\gamma_t, c(\gamma_j)) \tag{32}$$

where $J = \{t, t+1\}$ for $c(\gamma_t) = c_1$, $J = \{t-1, t, t+1\}$ for $c_1 < c(\gamma_t) < c_M$, and $J = \{t-1, t\}$ for $c(\gamma_t) = c_M$.

Expanding (32) we have

$$\psi_t = \begin{cases} b_{q_t}(o_t)\mu(\gamma_t, c(\gamma_t)) + b_{q_t}(o_{t+1})\mu(\gamma_t, c(\gamma_{t+1})) & : \; c(\gamma_t) = c_1 \\ b_{q_t}(o_{t-1})\mu(\gamma_t, c(\gamma_{t-1})) + b_{q_t}(o_t)\mu(\gamma_t, c(\gamma_t)) + b_{q_t}(o_{t+1})\mu(\gamma_t, c(\gamma_{t+1})) & : \; c_1 < c(\gamma_t) < c_M \\ b_{q_t}(o_{t-1})\mu(\gamma_t, c(\gamma_{t-1})) + b_{q_t}(o_t)\mu(\gamma_t, c(\gamma_t)) & : \; c(\gamma_t) = c_M \end{cases}$$

$$(33)$$

in which $c(\gamma_t)$ is the codeword to which γ_t is assigned.

Using logarithm, (31) can be expressed as

$$\log[\tilde{P}(O|Q, \lambda)] = \sum_{t=1}^{N} \log(\psi_t) \tag{34}$$

The unknown logo is then assigned to class κ^* as

$$\kappa^* = \arg\max_{\kappa}\{\log[\tilde{P}(O|Q, \lambda^\kappa)]\} \tag{35}$$

6 Logo Detection

Logo detection is formulated based on the principle that the spatial density of the foreground pixels within a given windowed image that contains a logo is greater than those of non-logo regions. We achieve this objective by considering each pixel as a potential cluster center of the windowed image and computing its spatial density function as follows. First we apply an image segmentation using a threshold algorithm such as Otsu's method [30] that is developed for grayscale images and outlined in Appendix B, to binarize the document image into foreground and background pixels. Let I be a document image of size $M \times N$, and $w \subset I$ a window of size $m \times n$, which is obtained as an approximation of a logo area, the spatial density function $M(p)$ with respect to the foreground pixels located at points $k \in w$, $1 \le k \le (m \times n)$, and p is the mid point of w acting as a cluster center can be determined by the mountain function [48]

$$M(p) = \sum_{k \in w, \, p \neq k} \exp[-\alpha d(p, k)], f_k = \text{foreground}, a < x(p) \le b, a < y(p) \le d$$

$$(36)$$

where α is a positive constant, $d(p, k)$ is a measure of distance between p and the foreground pixel located at k, $x(p)$ and $y(p)$ are the horizontal and vertical pixel coordinates of p respectively, $a = round(m/2)$, where $round(\cdot)$ is a round-off function, $b = M - round(m/2)$, $c = round(n/2)$, and $d = N - round(n/2)$.

A typical distance measure expressed in (36) is defined by

Table 1. Recognition rates of NN-based classifier

Subset	1	2	3	4	5	6
Logo Number	2-21	22-41	42-61	62-81	82-101	87-106
Noise	18	18	16	18	16	18
Total percentage	86.67 %					
Rotation (5 degrees)	16	15	13	14	15	17
Total percentage	75.00 %					

$$d(p,k) = [x(p) - x(k)]^2 + [y(p) - y(k)]^2 \tag{37}$$

The reason for using the mountain function instead of simply counting the number of the foreground pixels in the windowed region is that the foreground pixels of the logo region are more compact than those of non-logo regions. Therefore using (36), a region of pixels which are closely grouped together as a cluster tends to have greater spatial density than that of scattered pixels. For example, the number of the foreground pixels of a textual region can be the same or greater than those of the region having a fine logo; however using the mountain function, the results can be reversed with respect to the measure of spatial density. We will illustrate this effect in the experimental section by comparing between the mountain function defined in (36) and the counting of foreground pixels within a window w, denoted as $C(w)$ which is given by

$$C(w) = \sum_{k \in w} \delta(k) \tag{38}$$

where

$$\delta(k) = \begin{cases} 1 & : \quad f_k = \text{foreground} \\ 0 & : \quad f_k = \text{background} \end{cases} \tag{39}$$

Finally, the window w^* is detected as the region that contains a logo if

$$w^* = \arg\max_p M(p) \tag{40}$$

Furthermore, we can now see that the use of the mountain function $M(p)$ is preferred to the pixel-counting function $C(w)$ because the former, based on the focal point $p^* \in w^*$, can approximately locate the central pixel coordinates of a logo, which is then easily utilized to form a bounding box for clipping the whole detected logo.

In case the document page is a non-logo image, we can set up a threshold ξ by use of statistics to detect if there exists a logo:

288

Table 2. Recognition rates of pseudo HMM-based classifier

Degree of rotation	0	2	4	6	8	10
Without noise	103	103	100	101	95	95
Total percentage	94.76%					
With noise	98	99	98	92	89	88
Total percentage	89.52%					

Table 3. Recognition rates of FS-based classifier

Degree of rotation	0	2	4	6	8	10
Without noise	103	103	100	101	95	95
Total percentage	94.76%					
With noise	100	100	100	94	91	91
Total percentage	91.43%					

$$M(p^*) \begin{cases} > \xi & : \quad \mathcal{H}_1 \\ \leq \xi & : \quad \mathcal{H}_0 \end{cases} \tag{41}$$

where $\mathcal{H}_1$ is the hypothesis that a logo exists, and $\mathcal{H}_0$ the null hypothesis that a logo does not exist.

However, in practice false rejection is more of a concern than false acceptance [38] since a false rejection will miss out the detection and recognition of a logo, whereas a false acceptance can be corrected in the recognition phase if the output value is below a classification threshold.

7 Experiments on Logo Recognition

We use the logo database of the University of Maryland (UMD) [46] that consists of 105 intensity logo images of scanned binary versions (see Figures 7-13 in Appendix C) to test the proposed classifiers. For each image, the first 20 lags of the experimental semi-variograms are extracted to train the NN-based and the HMM-based classifiers.

A problem encountered with the NN approach is that it fails to simultaneously classify such a large number of classes that are 105 classes in this study. It has been observed that the networks become untrainable or poorly perform when the output classes/targets are approximately greater than twenty five. To proceed further with the study, the whole database is therefore divided into six subsets such that each subset consists of twenty logos. Thus, logos 1-20, 21-40, 41-60, 61-80, 81-100, and 86-105 are assigned to subset 1, subset 2, subset 3, subset 4, subset 5, and subset 6 respectively. For these partitions, similar logos may be grouped together thereby increasing the difficulty

Table 4. Detection rates for eight tests

Test number	1	2	3	4	5	6	7	8
Detection rate (%)	100	94.74	89.47	100	89.47	100	89.47	94.74
Total average rate (%)				94.74				

for the NN classifier. The networks are trained with the semi-variograms of the original images and are tested against images degraded with Gaussian noise of mean = 0 and variance = 0.02, and 5-degree rotation. Testing against noise, 18/20, 18/20, 16/20, 18/20, 16/20, and 18/20 are correctly recognized for subsets 1-6 respectively. This gives an overall recognition rate of 86.67%. Testing against a rotation of 5 degrees, 16/20, 15/20, 13/20, 14/20, 15/20, and 17/20 are correctly recognized for subsets 1-6 respectively. This gives an overall recognition rate of 75.00%. These results are also given in Table 1.

Each HMM with seven observation distributions is built for each logo. The observation symbols are generated by the LBG algorithm with a distortion threshold $\epsilon = 0.001$ (see Appendix A) These observation distributions include the original, 2-degree, 4-degree, 6- degree, 8-degree, 10-degree rotations, and Gaussian noise of zero mean and 0.02 variance. For the noise model, five simulations are carried out to construct the observation matrix. Testing for non-noisy logos, out of 105 logos, 103 (98.10%), 103 (98.10%), 100 (95.24%), 101 (96.19%), 95 (90.48%), and 95 (90.48%) are obtained for the original, 2-degree, 4-degree, 6-degree, 8-degree, and 10-degree rotations respectively, for which the total recognition rate is 94.76% . Testing against noisy images with one simulation, the results obtained are 98 (93.33%), 99 (94.29%), 98 (93.33%), 92 (87.62%), 89 (84.76%), and 88 (83.81%) for the original, 2- degree, 4-degree, 6-degree, 8-degree, and 10-degree rotations respectively, which yields the total recognition rate of 89.52%. These results are summarized in Table 2.

Using the same codewords generated for the pseudo HMM classifier, The fuzzy-set (FS) based classifier is applied to recognize the logos. For non-noisy logos, the recognition rates are 103 (98.10%), 103 (98.10%), 100 (95.24%), 101 (96.19%), 95 (90.48%), and 95 (90.48%) for the original, 2-degree, 4-degree, 6-degree, 8-degree, and 10-degree rotations respectively. Testing against noisy images with one simulation, the results obtained are 100 (95.24%), 100 (95.24%), 100 (95.24%), 94 (89.52%), 91 (86.67%), and 91 (86.67%) for the original, 2-degree, 4-degree, 6-degree, 8-degree, and 10-degree rotations respectively. These results are summarized in Table 3.

By the experimental results obtained from testing the three classifiers, we can see that neural networks are designed to approximate functions and therefore are capable of well recognizing patterns of similar functions such as the semi-variograms under study. The NN based classifier needs minimum training data and can predict unforseen patterns such as noise and rotated

Fig. 1. Skewed billing statement (left) and detected logo (right)

logos as presented in the cases studied above. However, it is limited to the classification in which a large number of classes involve. On the other hand, the so-called HMM based classifier is useful in classifying a large number of classes when given a reasonable amount of training data for likely encountered models. To overcome insufficient training data, we have built different models with deterministic and probabilistic observations where they are applicable. However, we have also degraded 105 original logos with Gaussian noise of zero mean and 0.01 variance and tested against the so-called HMM based classifier which correctly recognized 81 out of 105 logos (77.14%). This result, which is lower than the others presented above, is due to the fact that this noisy version was not included as training data in the observation distributions.

It can be seen that, in general, the pseudo HMM based classifier has achieved higher recognition rates than the NN based approach, and the fuzzy-set classifier obtains slightly higher results for the noisy patterns than the pseudo HMM based classifier.

8 Experiments on Logo Detection

The algorithm is tested with many logos, most of which are extracted from the public-domain UMD logo database [46], embedded in several document formats with respect to translation, scaling, orientation and degradation of the

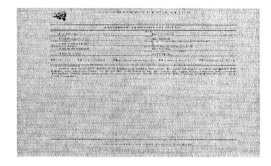

Fig. 2. Distorted faxed document (left) and detected logo (right)

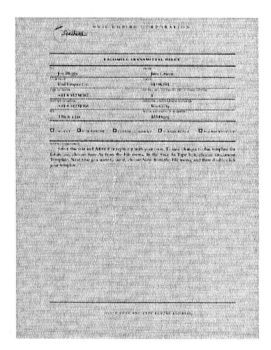

Fig. 3. Faxed document (left) and detected logo (right)

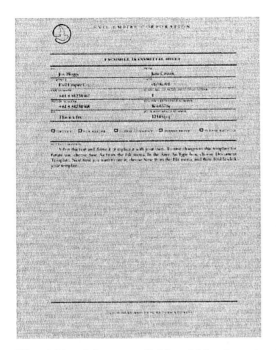

Fig. 4. Faxed document (left) and false detection (right) given by the pixel-counting method

logos. To increase the computational speed of the mountain function where $\alpha = 1$, $w = 21 \times 21$, the document image is reduced by dividing the original image five times coarser in both horizontal and vertical dimensions. Eight test models and the results are described as follows.

Test 1: Each of the 105 logos in the University of Maryland database is contained in a blank page of A4 size. The algorithm detects and extracts the logo using a 120-by-340 bounding box of which the center is the mid point of the detected windowed image. The detection rate is 100%.

Test 2: Including the presence of several single logos, each contained on 38 textual pages in the forms of letter, faxed document, and billing statement. The detection rate is 94.74%.

Test 3: Several different single logos, each contained on 38 textual pages in the forms of letter, faxed document, and billing statement with skewing to the left at specific angles. The detection rate is 89.47%.

Test 4: Fifty logos, each contained on a blank page with skewing to the left at specific angles are to be detected. The detection rate obtained is 100%.

Test 5: Different single logos, each contained on 38 textual pages in the forms of letter, faxed document, and billing statement with skewing to the left at specific angles. The detection rate is 89.47%.

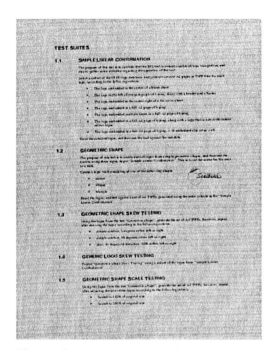

Fig. 5. Letter document (left) and false detection (right) for Logo 103 given by both pixel-counting and mountain functions

Test 6: Fifty logos each of which is contained on a blank page with skewing to the right at specific angles. The detection rate obtained is 100%.

Test 7: Several logos each of which is contained on 38 textual pages in the forms of letter, faxed document, and billing statement, are geometrically changed by distortion, scaling and rotation including up-side- down. The detection rate is 89.47%.

Test 8: The set of document images used in Test 2 is degraded with Gaussian white noise of zero mean, and 0.001 variance. The detection rate is 94.74%.

The results of these tests are summarized in Table 4. Figure 1 and Figure 2 show a skewed billing statement, and a distorted faxed document with the corresponding detected logos respectively. Figure 3 shows a faxed document and Logo 103 as the correctly detected logo, whereas Figure 4 illustrates the false detection using the pixel counting function $C(w)$ as a textual region instead of correctly detecting Logo 17 (the dove) which is then correctly detected by the mountain function $M(p)$. It is observed that both pixel-counting and mountain functions particularly fail to detect Logo 103 contained on an A4-size letter in tests 3, 5, 7, and 8 with false detections as textual regions as shown in Figure 5 for test 3. The reason for this is explained by the fact that the pixel density of this logo, that mostly consists of thin curves, is less (not distinctive) than those of some rich textual areas. A faxed document degraded

Fig. 6. Degraded document (left) and detected logo (right)

with Gaussian white noise (0 mean, 0.001 variance) and the detected logo are shown in Figure 6.

9 Concluding Remarks

The UMD logo database we have studied herein is smaller in size in comparison with other image databases for logo/trademark retrieval; however, the task of logo recognition requires more rigorous criteria in the scoring procedure. Furthermore, the intensity levels of these logos are almost binary, which leave little information for feature extraction with respect to other intensity or textural images. To tackle this problem, we have introduced a geostatistical method for extracting the spatial features of the logos, which prove useful for training the classifiers. In case of small classes, the NN-based classifier is more convenient to implement. When the number of classes increases, the pseudo HMM-based classifier can solve the problem more effectively, but it requires a good amount of training data for handling noisy patterns. The FS-based classifier can slightly improve the results by relaxing the hard partitions of the VQ codewords used for the pseudo HMM-based classifier. As the extracted features and the classifiers work well for the logo recognition problem, they have the potential to be applied to the retrieval of document images.

Our proposed algorithm for logo detection based on the mountain function is simple, effective, and reasonably fast for real-time applications. However, we do not consider the inclusion of pictures in document images, which is a further issue for our future research.

Acknowledgments: The author would like to thank his Head of Group, Dr. Richard Price, for many helpful discussions, and his continuing encouragement on the author's research in image document analysis. The faxed document images are provided by the Image Research Laboratory of the Queensland University of Technology, Brisbane, Australia.

References

1. C.M. Bishop, *Neural Networks for Pattern Recognition.* Oxford University Press, New York, 1995.
2. Z. Chi, H. Yan, and T. Pham, *Fuzzy Algorithms: With Applications to Image Processing and Pattern Recognition.* World Scientific, Singapore, 1996.
3. V. Chandran, B. Carswell, S.L. Elgar, and B. Boashash, Pattern recognition using invariants defined from higher order spectra: 2-D image inputs, *IEEE Trans. Signal Processing,* 6 (1997) 703-712.
4. M.T. Chang, and S.Y. Chen, Deformed trademark retrieval based on 2D pseudo-hidden Markov model, *Pattern Recognition,* 34 (2001) 953-967.
5. H.D. Cheng, C.Y. Wu, and D.L. Hung, VLSI for moment computation and its application to breast cancer detection, *Pattern Recognition,* 31:9 (1998) 1391-1406.
6. G. Ciocca, and R. Schettini, Content-based similarity retrieval of trademarks using relevance feedback, *Pattern Recognition,* 34 (2001) 1639-1655.
7. G. Cortelazzo and G. A. Mian and G. Vezzi and P. Zamperoni, Trademark shapes description by string-matching techniques, *Pattern Recognition,* 27:8 (1994) 1005-1018.
8. M. David, *Geostatistical Ore Reserve Estimation.* Elsevier, Amsterdam, 1977.
9. D.S. Doermann, E. Rivlin, and I. Weiss, Logo recognition using geometric invariants, *Int. Conf. Document Analsysis and Recognition,* pp. 894-897, 1993.
10. D.S. Doermann, E. Rivlin, and I. Weiss, Logo Recognition, *Technical Report: CS-TR-3145,* University of Maryland, 1993.
11. C.V. Deutsch, and A.G. Journel, *GSLIB: Geostatistical Software Library and User's Guide.* Oxford University Press, 2nd edition, New York, 1998.
12. S.A. Dudani, K.J. Kenneth, and R.B. McGhee, Aircraft identification by moment invariants, *IEEE Trans. Computers,* 26 (1977) 39-46.
13. C.S. Fuh, S.W. Cho, and K. Essig, Hierarchical color image region segmentation for content-based image retrieval system, *IEEE Trans. Image Processing,* 9:1 (2000) 156-163.
14. B. Gunsel, and A.M. Tekalp, Shape similarity matching for query-by-example, *Pattern Recognition,* 31:7 (1998) 931-944.
15. M.E. Hohn, *Geostatistics and Petroleum Geology.* Van Nostrand Reinhold, New York, 1988.
16. M.K. Hu, Pattern recognition by moment invariants, *Proc. IRE,* 49 (1961) 1428.

17. M.K. Hu, Visual pattern recognition by moment invariants, *IRE Trans. Information Theory*, IT-8 (1962) 179-187.
18. E.H. Isaaks, and R.M. Srivastava, Spatial continuity measures for probabilistic and deterministic geostatistics, *Mathematical Geology*, 20:4 (1988) 313-341.
19. E.H. Isaaks, and R.M. Srivastava, *An Introduction to Applied Geostatistics*. Oxford University Press, New York, 1989.
20. A.K. Jain, and A. Vailaya, Shape-based retrieval: A case study with trademark image databases, *Pattern Recognition*, 31 (1998) 1369-1390.
21. A.G. Journel, and Ch.J. Huijbregts. *Mining Geostatistics*. Academic Press, Chicago, 1978.
22. A. Khotanzad, and Y.H. Hong, Invariant image recognition by Zernike moments, *IEEE Trans. Pattern Analysis & Machine Intelligence*, 12:5 (1990) 489-497.
23. Y.C. Kim, and E.J. Powers, Digital bispectral analysis and its applications to nonlinear wave interactions, *IEEE Trans. Plasma Science*, 7 (1979) 120-131.
24. H.K. Lee, and S.I. Yoo, Intelligent image retrieval using neural network, *IEICE Trans. Information and Systems* E84-D:12 (2001) 1810-1819.
25. Y. Linde, A. Buzo, and R.M. Gray, An algorithm for vector quantizer design, *IEEE Trans. Communications*, 28:1 (1980) 84-95.
26. S. Loncaric, A survey of shape analysis techniques, *Pattern Recognition*, 31:8 (1998) 983-1001.
27. G. Matheron, La theorie des variables regionalisees et ses applications, *Cahier du Centre de Morphologie Mathematique de Fontainebleau*, Ecole des Mines, Paris, 1970.
28. D.A. Mitzias, and B.G. Mertzios, Shape recognition with a neural classifier based on a fast polygon approximation technique, *Pattern Recognition*, 27:5 (1994) 627-636.
29. G. Nagy, and S. Seth, Hierarchical representation of optical scanned documents, *Proc. Seventh Int. Conf. Pattern Recognition*, vol. 1, pp. 347-349, 1984.
30. N. Otsu, A threshold selection method from gray-level histograms, *IEEE Trans. Systems, Man, and Cybernetics*, 9:1 (1979) 62-66.
31. H.L. Peng, and S.Y. Chen, Trademark shape recognition using closed contours, *Pattern Recognition Letters*, 18 (1997) 791-803.
32. J.R. Quinlan, *C4.5: Programs for Machine Learning*. Morgan Kauffmann, San Mateo, California, 1992.
33. L.R. Rabiner, A tutorial on hidden Markov models and selected applications in speech recognition, *Proc. IEEE*, 77:2, 257-286 (1989).
34. L.R. Rabiner, and B.H. Juang, *Fundamentals of Speech Recognition*. Prentice-Hall, New Jersey (1993).
35. G. Ramstein, and M. Raffy, Analysis of the structure of radiometric remotely-sensed images, *Int. J. Remote Sensing*, 10 (1989) 1049-1073.
36. Y. Rui, T.S. Huang, and S. Chang, Image retrieval: Current techniques, promising directions, and open issues, *J. Visual Communication and Image Representation*, 10 (1999) 39-62.
37. S.S. Sarkaria, and A.J. Harget, Shape recognition using Kohonen self-organising feature map, *Pattern Recognition Letters*, 13 (1992) 189-194.
38. S. Seiden, M. Dillencourt, S. Irani, R. Borrey, and T. Murphy, Logo detection in document images, *Proc. Int. Conf. Imaging Science, Systems, and Technology*, pp. 446-449, 1997.

39. A.W.M. Smeulders, M. Worring, S. Santini, A. Gupta, and R. Jain, Content-based image retrieval at the end of the early years, *IEEE Trans. Pattern Analysis & Machine Intelligence* 22:12 (2000) 1349-1380.

40. Y. Shao, and M. Celenk, Higher-order spectra (HOS) invariants for shape recognition, *Pattern Recognition*, 34 (2001) 2097-2113.

41. R. Srivastava, and H. Parker, Robust measures of spatial continuity, *Third Int. Geostatistics Congress* (M. Armstrong et al, ed.) Dordrecht, Holland, 1988.

42. J.-L. Starck, F. Murtagh, and A. Bijaoui. *Image Processing and Data Analysis: The Multiscale Approach.* Cambridge University Press, Cambridge, UK, 1998.

43. Y.Y. Tang, E.C.M. Lam, New method for feature extraction based on fractal behavior, *Pattern Recognition*, 35 (2002) 1071-1081.

44. C.H. Teh, and R.T. Chin, Image analysis by the methods of moments, *IEEE Trans. Pattern Analysis and Machine Intelligence*, 10:4 (1988) 496-513.

45. H.P. Tseng, M.J. Sabin, and E.A. Lee, Fuzzy vector quantization applied to hidden Markov modeling, *Proc. IEEE ICASSP*, vol. 2, pp. 641-644, 1997.

46. ftp://ftp.cfar.umd.edu/pub/documents/contrib/databases/UMDlogo_database.tar

47. J. Wood, Invariant pattern recognition: A review, *Pattern Recognition*, 29:1 (1996) 1-17.

48. R.R. Yager and D.P. Filev, Approximate clustering via the mountain method, *IEEE Trans. Systems, Man, and Cybernetics*, 24 (1994) 1279-1284.

49. H.W. Yoo, S.H. Jung, D.S. Jang, and Y.K. Na, Extraction of major object features using VQ clustering for content-based image retrieval, *Pattern Recognition*, 35 (2002) 1115-1126.

50. C. Yuceer, and K. Oflazer, A rotation, scaling, and translation invariant pattern classification system, *Pattern Recognition*, 26:5 (1993) 687-710.

51. L.A. Zadeh, Probability measures of fuzzy events, *J. Mathematical Analysis and Applications*, 23 (1968) 421-427.

APPENDIX A – *LBG Algorithm*

1. Given a training data set $\mathbf{X} = \{\mathbf{x}_1, \mathbf{x}_2, \ldots, \mathbf{x}_M\}$, where $\mathbf{x}_m = (x_{m1}, x_{m2}, \ldots, x_{mk})$; $m = 1, 2, \ldots, M$.
2. Given $\epsilon > 0$ (small real number)
3. Set $N = 1$, compute initial cluster center and average distortion

$$\mathbf{c}_1^* = \frac{1}{M} \sum_{m=1}^{M} \mathbf{x}_m \tag{42}$$

$$D^* = \frac{1}{Mk} \sum_{m=1}^{M} \|\mathbf{x}_m - \mathbf{c}_1^*\|^2 \tag{43}$$

4. Splitting:

$$\mathbf{c}_{i1} = (1 + \epsilon)\mathbf{c}_i^*, \ 1 \le i \le N$$
$$\mathbf{c}_{i2} = (1 - \epsilon)\mathbf{c}_i^*, \ 1 \le i \le N$$

Set $N = 2N$

5. Iteration: Set $j = 0$ and let $D^j = D^*$

   a) Assign vector to closest codeword

$$V(\mathbf{x}_m) = \mathbf{c}_n^* = \arg\min_n \|\mathbf{x}_m - \mathbf{c}_n^j\|^2, \ 1 \le m \le M, 1 \le n \le N \tag{44}$$

   b) Update cluster centers

$$\mathbf{c}_n^{j+1} = \frac{1}{N(V)} \sum_{V(\mathbf{x}_m)=\mathbf{c}_n^*} \mathbf{x}_m, \ 1 \le n \le N \tag{45}$$

   where $N(V)$ is the number of $V(\mathbf{x}_m) = \mathbf{c}_m^*$.

   c) Set $j = j + 1$, and compute

$$D^j = \frac{1}{Mk} \sum_{m=1}^{M} \|\mathbf{x}_m - V(\mathbf{x}_m)\|^2 \tag{46}$$

   d) Go to step (*a*) if

$$\frac{D^{j+1} - D^j}{D^{j+1}} > \epsilon \tag{47}$$

   e) Set $D^* = D^j$, and relabel $\mathbf{c}_n^* = \mathbf{c}_n^j, \ 1 \le n \le N$

6. Repeat steps 4 and 5 until the desired number of codewords is obtained.

APPENDIX B – *Segmentation using Otsu's Threshold Selection*

Segmentation of an intensity image using Otsu's algorithm [30] can be outlined as folllows [2].

Let $h(z)$ be the normalized histogram function that represents the percentage of pixels $f(x,y)$ having a gray level $z \in [0, L-1]$ over the total number of pixels in the image.

The objective function $J(T)$ as a measure of class separability is to be maximized:

$$J(T) = \frac{P_1(T)P_2(T)[\mu_1(T) - \mu_2(T)]^2}{P_1(T)\sigma_1^2(T) + P_2(T)\sigma_2^2(T)} \tag{48}$$

where

$$P_1(T) = \sum_{z=0}^{T} h(z) \tag{49}$$

$$P_2(T) = 1 - P_1(T) \tag{50}$$

$$\mu_1(T) = \frac{1}{P_1} \sum_{z=0}^{T} zh(z) \tag{51}$$

$$\mu_2(T) = \frac{1}{P_2} \sum_{z=T+1}^{L-1} zh(z) \tag{52}$$

$$\sigma_1^2(T) = \frac{1}{P_1} \sum_{z=0}^{T} [z - \mu_1(T)]^2 h(z) \tag{53}$$

$$\sigma_2^2(T) = \frac{1}{P_2} \sum_{z=T+1}^{L-1} [z - \mu_2(T)]^2 h(z) \tag{54}$$

The optimal threshold T^* is selected by

$$T^* = \arg \max_{0 \le T \le L-1} J(T) \tag{55}$$

The image $f(x,y)$ is then assigned as either a foreground or a background pixel by the following criteria:

$$f(x,y) \begin{cases} > T^* & = \text{foreground} \\ \le T^* & = \text{background} \end{cases} \tag{56}$$

Fig. 7. Logo numbers 2-16

Fig. 8. Logo numbers 17-31

Fig. 9. Logo numbers 32-46

Fig. 10. Logo numbers 47-61

Fig. 11. Logo numbers 62-76

305

Fig. 12. Logo numbers 77-91

306

Fig. 13. Logo numbers 92-106

Fuzzy Web Information Classification Agents

Yanfei Wang[1] and Yan-Qing Zhang[1]

Department of Computer Science, Georgia State University, Atlanta, GA 30303, USA. yanfei_wang@yahoo.com and yzhang@cs.gsu.edu

1 INTRODUCTION

We are in an era with information overload. It becomes more and more difficult for us to analyze the fast-changing information and make decision upon the analyses. The web information agent appears to give us an easy way. In general, intelligent Web agents based on the CWI (Computational Web Intelligence) techniques can help a better e-Business [15]. CWI is a hybrid technology of Computational Intelligence (CI) and Web Technology (WT) dedicating to increasing QoI (Quality of Intelligence) of e-Business applications on the Internet and wireless networks [15]. Fuzzy computing, neural computing, evolutionary computing, probabilistic computing, granular computing, rough computing, WT, data mining, personalization and intelligent agent technology are major techniques of CWI. Currently, seven major research areas of CWI are (1) Fuzzy WI (FWI), (2) Neural WI (NWI), (3) Evolutionary WI (EWI), (4) Probabilistic WI (PWI), (5) Granular WI (GWI), (6) Rough WI (RWI), and (7) Hybrid WI (HWI). In the future, more CWI research areas will be added. Here, FWI techniques are mainly used in the fuzzy Web information classification agent.

An agent is defined as "generally, a person or thing that acts or is capable of acting, or, in this comparison, one who or that which acts, or is empowered to act, for another.[5]" Generally in the scope of data mining, agent stands for some software that can perform some kind of automated actions. There are several variations of the definition of agents. Here is one from IBM: Intelligent agents are software entities that carry out some set of operations on behalf of a user or another program with some degree of independence or autonomy and in so doing, employ some knowledge or representation of the user's goals or desires".

The reliable and accurate results are measured by comparing the information web agent offered versus actual one. So the algorithm is critical. The good algorithm gives us the successful agent. In the real world, lots of things are uncertainties that people uses "probably" more common than critical point such as "Yes" or "No". Fuzzy logic technology can generate decisions to reduce the risk by using approximate information.

Fuzzy logic is a superset of conventional (Boolean) logic that has been extended to handle the concept of partially true-partially false values between "complete truth" and "complete false" [6,11,12].

It is a well-known fact that tracking changes of thousands of daily stocks is very difficult job for humans to perform. On the other hand, automating the process by an intelligent stock agent running on a computer can perform this task and help human stock traders or a stock agent a lot in terms of time saving, reliability, efficiency and convenience [13]. In large, the intelligent agent technology and soft computing techniques are useful for smart Web applications under uncertainty conditions [6,13].

Personalized agents have merits including trust (the agent will do what the user wants, personalization (the agent can either learn or be explicitly taught what to do for each individual user), and autonomy (the agent is allowed to take at least some actions on the user's behalf, without permission or perhaps even notification).

By including uncertainties, fuzzy reasoning can yield satisfactory results comparable to human reasoning. For data mining applications [1,10], fuzzy reasoning is easy to understand and implement with the help of some popular complex techniques used in data mining systems [3,14]. In addition, fuzzy reasoning uses fuzzy rule base for input classification and processing which enforces the precision and reliability of the output. Fuzzy systems generally enhance users confidence and trust on the results.

The Web-based stock information agent using fuzzy logic is implemented to get the latest stock price information. Fuzzy logic implementation becomes very complex when the number of inputs increases. Thus, a minimum number of inputs should be carefully selected in order to avoid system complexity and overhead while provide enough precision in output. Here we choose "earning/share", "P/E Ratio" as inputs in the present system implementation among the so many values that can be fetched from Internet. The output of the fuzzy logic method is in the form of a list of predicted stock value. The method is processing information based on 25 rules that takes the before mentioned factors into account. Also it is flexible and easy to adjust the rules or input parameters in fuzzy logic according to actual condition.

The web-based application - fuzzy web stock information agent - "VirtualStreet.com" is created based on Jakarta-Tomcat-3.2.1 (web sever) and Microsoft SQL Server 7.0 (database). A variety of languages, including Java, JSP (Java Sever Page), HTML (Hyper-Text Makeup Language), Java Servlet, XML, XSL are used. The agent enables the users to create their own portfolios that contain the watch list of stocks. Using fuzzy reasoning, the Agent can create a list of top 10 stocks based on output values calculated from stock information [8,9].

2 Intelligent Web Agent based on the FWI

2.1 Problem Specification

In today's fast-paced, on-line business environment, a lot of important information needs people to make correct decision. Today, leveraging the investment in your information or data by analyzing and using the knowledge is critical for your business to archive and maintain its competitive edge. For example, when one wants to do some business, he/she needs to make decision on what to sell, how to sell, how to market, who like this products, etc. How to make those decisions correctly? The answer is that he or she can find the solution from various sources of data. People store more and more data years after years. And we are facing with an explosion in the amount of data that must be stored and processed. Fortunately, cheap storage makes data storage much easy than before. But decision-making is becoming more and more difficult. People may get some information according to their experience or gut feeling from the history data.

It is clear that it would be incredibly difficult for a human to fully analyze all the potential relationships among the tremendous amount of data and information. What's more, these relationships are subject to change weekly, or daily, or even hourly! So a rapid analysis of data is virtually unfeasible for a person to handle. But it is very important to master them in today's life. This need has pushed the industry to develop Data Mining specialists and Data Mining techniques, often particular to a particular need.

Although we can analyze data and make decision by professional experience or intuition feeling, the formal research is more accurate. Why? The reason is simple. Although experience can provide the results, it may be obsolete to analyze the data correctly for the current situation. Intuition feeling may give the result that is not always true. It is random to guess. Finally, research is a formal method of gathering data and analyzing it. So whether we like it or not, the business world is made up of individuals making business decisions as really, just more informed guesses.

Today, everything changes so quickly. If one fails to pay attention to the trend of the market now, he/she can be a loser soon. Many problems, one solution is Intelligent Web Agents based on the CWI.

2.2 Intelligent FWI Web Agent

FWI

FWI has two major techniques which are (1) fuzzy logic and (2) WT. The main goal of FWI is to design intelligent fuzzy e-agents which can deal with fuzziness of data, information and knowledge, and also make satisfactory decisions like the human brain for e-Business applications effectively.

Intelligent Web Agent

We interact with agents in our daily life to perform some specific tasks on behalf of us and fulfill our needs. When we plan to travel, travel agents will help us make trip travel, buy air tickets, and reserve hotels. Thinking about buying a home to be a first-time homeowner? Chances are that you may not be experienced at first time. A real estate agent can work with you through the home buying 101 steps. Agents take what we need or our requirements as "input", and along the information they know, and process all these data, and finally come out with results or recommendations.

An agent can be defined as one that acts or exerts power. Comparing to old fashion person agents, "soft-agents", agent-based technologies begin to play a key role in assisting our daily life. In fact, an agent, a.k.a. software robot, is a software program that performs tasks for its user using artificial intelligence. It employs automated and efficient ways to find information that the user wants, and acts on it (autonomy and personalization merits). It can think and will act on behalf of a user to carry out tasks (trust merit).

3 FUZZY LOGIC SYSTEM

Most often we face real-world vagueness, since the world we live in and natural language we use abound with vague and less-precise concepts. We have to make decision upon imperfect collection of information. But the requirements of making good decision upon imprecise and fast-changing information already exceed the capability of human brain. When we begin to reply on software agent to process massive data and further take over decision-making process, we want it to be intelligent, and to some degree to mimic human decision-making.

An intelligent agent is a software programs that implements artificial intelligent technique. The algorithm plays a key role here. In the daily life many things are more important than others, based on relative relationships. It is difficult to solve every factor in precise percentage among all the factors. But people can decide which one is more important than others. So we can use imprecise rather than precise. Fuzzy rule system provides such a way. It can solve the problem using approximate information and uncertainty to make decisions.

A fuzzy reasoning with fuzzifier and defuzzifier is one that does not use Boolean logic, but using fuzzy logic. Introduced by Zadeh in 1960's. Fuzzy logic is a superset of conventional (Boolean) logic that has been extended to handle the concept of partial truth - truth values between "complete falseness" and "complete truth". Unlike the process of traditional logic which requires a clear understanding of a system, usually with exact equations and precise numeric values, fuzzy logic incorporates an alternative way of thinking, which uses abstraction of the imprecise, and translates subjective concepts such as very old into exact numeric ranges (that describe to what degree the guiding rule is applicable). The notion central to fuzzy system is that truth values (in fuzzy logic) or membership values (in fuzzy sets) are indicated by a value on the range of [0.0, 1.0], with 0.0 representing "complete falseness" and 1.0 representing "complete truth".

We get used to think under imprecise and uncertain conditions. For example, let's say the age. We say someone is young. But how old is young? When John is 20, we can consider him young. But if James is 12, obviously John is older than James. By setting a young fuzzy subset by a membership function, we can easily to know what degree of young is based on a person's age.

A fuzzy rule system is a collection of membership functions and rules that are used to reason about data. Each rule defines a fuzzy implication that performs a mapping from fuzzy input states set to the corresponding fuzzy output value. The goal is to get the numerical process using fuzzy rule systems. Generally, The rules in a fuzzy rule system are similar to the following form:

If x is low and y is very low then z is very low.

where x and y are input linguistic variables, z is an output linguistic variable. low is a a membership function defined on x and y are defined as, very low is membership function that is defined on output variable z, the part between "if" and "then" is a fuzzy logic expression that describes to what degree the rule is applicable. The rest part of the rule following "then" assigns a membership function to output variable.

Generally, fuzzy logic system includes fuzzy rule base and membership functions, fuzzy inference engine, fuzzifier, and defuzzifier. The relationships are shown in Fig. 1.

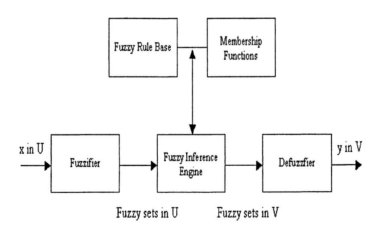

Fig. 1. Basic configuration of a fuzzy logic system

A fuzzy rule base includes a collection of conditional statements in the form of if-then rules. Examples of these fuzzy rules are:

Rule (1): If x is low and y is low then z is high.

Rule (2):If x is low and y is medium then z is medium.

..

Rule (n): if x is high and y is high then z is low.

where U Rule (n), and V one of Rule (l).

Fuzzy interference engine rely on membership functions to calculate the correct value between 0 and 1. The degree to which any fuzzy rule is true is denoted by a numeric value between 0 and 1. Fuzzifier performs a mapping from a crisp point to a fuzzy set. Defuzzifier performs the opposite operation.

Fig. 2 shows a fuzzy set that is characterized by a membership function, $U \in [0, 1]$. The membership function has three fuzzy sets: low, medium, and high. It should be noted that since we do not need to specify a precise border between the ranges of input values and each input value has a membership value, for a crisp point, it may fall in two adjacent fuzzy sets, for example, in both low and medium sets, in Fig. 2.

A typical fuzzy reasoning includes four processes: fuzzification, inference, composition and defuzzification. Fuzzifier accepts input values (in the form of crisp points) to be mapped to fuzzy sets. Fuzzy inference engine performs prescribed calculations, and then dufuzzifier maps fuzzy sets to crisp points using numerical method.

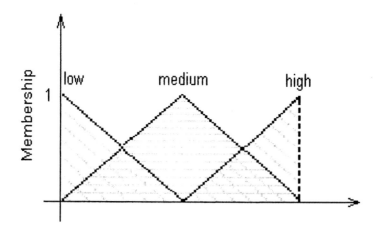

Fig. 2. A typical membership function with three fuzzy sets

In the fuzzification process, the membership functions defined on the input variables are applied to their actual (crisp) values, to determine the degree of truth for each rule premise. If a rule's premise has a nonzero degree of truth, then the rule is referred as fire up. In the inference process, the entire set of rules is evaluated, and the fuzzy output truth value is computed for the premise of each rule, and applied to the conclusion part of each rule. In the composition process, all of the fuzzy subsets that are assigned to each output variable are combined into a single fuzzy subset for each output variable. In the defuzzification process, the fuzzy output value is converted back into a crisp value. Two main methods of defuzzification are CENTROID and MAXIMUM. In CENTROID method, the crisp value of the output valuable is calculated by finding the variable value of the center of gravity of the membership function for the fuzzy value, while in MAXIMUM method one of the variable values at which the fuzzy subset has its maximum truth values is chosen as the crisp value for the output variable.

It is important to select proper linguistic values and rules in the rule base. Such selection is usually based on a combination of human expertise and trial-and-error processes on the particular situation. There exists tradeoff to keep a minimum number of linguistic rules to make system simple and efficient, while still maintaining the desired accuracy to achieve an acceptable goal or performance. Fuzzy logic resembles human decision-making with its ability to find precise solutions from approximate information. So, fuzzy logic systems affords a broader, richer field of information and the manipulation of the information than do traditional methods. While fuzzy logic is a paradigm compared to conventional methodology, it also offers additional benefits, including simplicity and time saving. Fuzzy logic provides a remarkably simple way to draw definite conclusions from vague, ambiguous or imprecise information using linguistic rules. It does not require complex quantification and math equations governing the mapping between inputs and outputs. With a fuzzy logic design, some time-consuming steps (such as quantifying, modeling, and programming) are eliminated. And since fuzzy logic is rule based, time can be saved by focusing on application development instead of extensive high- or low-level programming. As a result, fuzzy logic implementation reduces the overall design cycle.

Today, fuzzy logic finds its way in a variety of control applications including chemical process control, manufacturing, and pattern recognition.

4 FUZZY WEB STOCK INFORMATION AGENT

A small stock information application - fuzzy web stock information agent VirtualStreet - will be designed to demonstrate how can intelligent web agent works, or helps decision-making process. Usually several factors (considered as input values) contribute to the decision of which stock to buy or sell (output of decision-making). But it is difficult to say one factor is much more important than the others, for example, it may not be appropriate to say ask price overweighs earning per share or earning per share overweighs price per earning. Such a decision-making situation is which fuzzy logic method is capable to handle, given that each input factor can be characterized by a value range.

The advantage of fuzzy logic is that we do not need to specify a precise border between the ranges of input values and each input value has a membership value. In order to simplify the system, only two values are considered in the designed application as inputs, which are earning per share (E) and price per earning (P/E). The output will be the evaluation of chances that the stock is recommended as "good" and worth to invest, a decision made by the agent on behalf of the investor (or user). System design of the fuzzy logic personalized stock agent is described in detail as follows.

Step 1: Design a fuzzy logic control system

A fuzzy logic control system with two inputs and one output is shown in Fig. 3. The input are earning per share (E) and price per earning (P/E). The output is a value (S) that shows the degree of membership within the set of the stocks that are recommended as good.

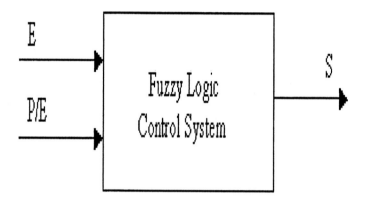

Fig. 3. A fuzzy logic control system with two inputs and one output

Step 2: Create the fuzzy rule base

The fuzzy partition of E includes Very Low (VL), Low (L), Middle (M), High (H), Very High (VH). The corresponding membership function is shown in Fig. 4, with a range of earning/share from 0 to 5.

The fuzzy partition of P/E includes Very Low (VL), Low (L), Middle (M), High (H), Very High (VH). The corresponding membership function is shown in Fig. 5, with a range of P/E value from 0 to 80 for most stocks.

The fuzzy partition of S also includes Very Low (VL), Low (L), Middle (M), High (H), and Very High (VH), as shown in Fig. 6.

Then a rule base that consists of 25 rules can be built according to the method mentioned before. The 25 fuzzy rule base are given below:

Rule 1. If earning is very low and P/E is very low, then the degree of the stock's membership within "good stocks" is low.

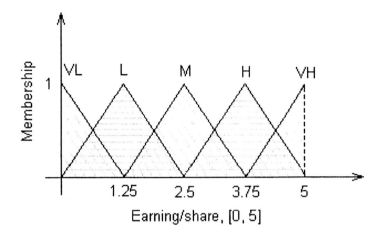

Fig. 4. Earning/share membership function

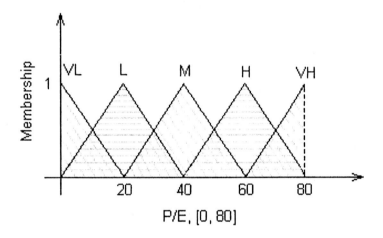

Fig. 5. P/E membership function

Rule 2. If earning is low and P/E is very low, then the degree of the stock's membership within "good stocks" is medium.

Rule 3. If earning is medium and P/E is very low, then the degree of the stock's membership within "good stocks" is medium.

Rule 4. If earning is high and P/E is very low, then the degree of the stock's membership within "good stocks" is very high.

Rule 5. If earning is very high and P/E is very low, then the degree of the stock's membership within "good stocks" is very high.

Rule 6. If earning is very low and P/E is very low, then the degree of the stock's membership within "good stocks" is very low.

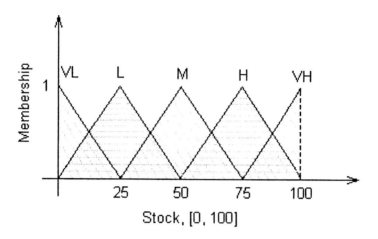

Fig. 6. Stock evaluation membership function

Rule 7. If earning is low and P/E is low, then the degree of the stock's membership within "good stocks" is low.

Rule 8. If earning is medium and P/E is low, then the degree of the stock's membership within "good stocks" is very medium.

Rule 9. If earning is high and P/E is low, then the degree of the stock's membership within "good stocks" is high.

Rule 10. If earning is very high and P/E is low, then the degree of the stock's membership within "good stocks" is very high.

Rule 11. If earning is very low and P/E is medium, then the degree of the stock's membership within "good stocks" is very low.

Rule 12. If earning is low and P/E is medium, then the degree of the stock's membership within "good stocks" is very low.

Rule 13. If earning is medium and P/E is medium, then the degree of the stock's membership within "good stocks" is low.

Rule 14. If earning is high and P/E is medium, then the degree of the stock's membership within "good stocks" is medium.

Rule 15. If earning is very high and P/E is medium, then the degree of the stock's membership within "good stocks" is high.

Rule 16. If earning is very low and P/E is high, then the degree of the stock's membership within "good stocks" is very low.

Rule 17. If earning is low and P/E is high, then the degree of the stock's membership within "good stocks" is very low.

Rule 18. If earning is medium and P/E is high, then the degree of the stock's membership within "good stocks" is very low.

Rule 19. If earning is high and P/E is high, then the degree of the stock's membership within "good stocks" is low.

Rule 20. If earning is very high and P/E is high, then the degree of the stock's membership within "good stocks" is medium.

Rule 21. If earning is very low and P/E is very high, then the degree of the stock's membership within "good stocks" is very low.

Rule 22. If earning is low and P/E is very high, then the degree of the stock's membership within "good stocks" is very low.

Rule 23. If earning is medium and P/E is very high, then the degree of the stock's membership within "good stocks" is very low.

Rule 24. If earning is high and P/E is very high, then the degree of the stock's membership within "good stocks" is very low.

Rule 25. If earning is very high and P/E is very high, then the degree of the stock's membership within "good stocks" is low.

Step 3: Fuzzification

In this step, the actual values are used in the membership functions defined on the input variables to determine the degree of truth for each rule promise. For example, assuming Intel's earning/share is 0.78 and P/E is 27.79 at some particular time. According to the fuzzy partitions in Fig. 7, we can know that when earning/share = 0.78, VeryLow(earning/share)=0.376, and Low(earning/share)=0.624.

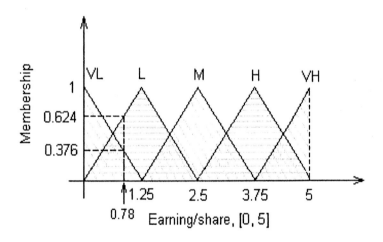

Fig. 7. Earning/share membership function

Similarly, according to the fuzzy partitions in Fig. 8, we can know that when P/E=27.79, Low(P/E)=0.6105, and Medium(P/E)=0.3895.

Four fuzzy rules (Rules 1, 7, 11 and 12) are fired.

Step 4: Defuzzification

Here we choose CENTROID method in which the center of gravity is used to determine the weight from the output membership values, as shown in Fig. 9.

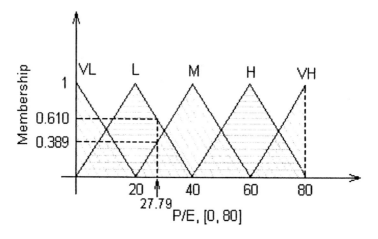

Fig. 8. P/E Membership Function

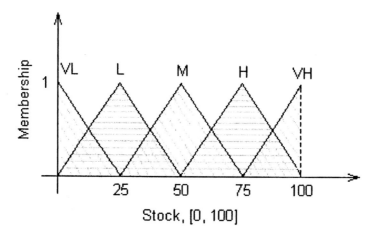

Fig. 9. Evaluation stock membership function

In this case: When earning/share is very low and P/E is low, the weight is determined from the first output triangle, representing that the weight is equal to 25/3=8.33.

In the second fired rule, when earning is low and P/E is medium, the weight is determined from the first output triangle, and is equal to 25/3=8.33.

In the third fired rule, when earning is very low and P/E is medium, the weight is determined from the first output triangle, and is equal to 25/3=8.33.

In the forth fired rule, when earning is low and P/E is low, the weight is determined from the second output triangle, and is equal to 25.

The final crisp value of the output score will be calculated by using the Average of Gravities (AOG) method:

$$Y = \frac{\sum_{k=1}^{m} G_k \mu_{A_1^k}(x_1) \mu_{B_1^k}(x_2)}{\sum_{k=1}^{m} (\mu_{A_1^k}(x_1) \mu_{B_1^k}(x_2))}, \tag{1}$$

where G_k is the value of the center of gravity of the membership function, and Y (evaluated stock in this case) is output, x_1 and x_2 are inputs (in this case, are earning/share and P/E, respectively). $\mu_{A_1^k}(x_1)$ and $\mu_{B_1^k}(x_2)$ represent the membership functions of x_1 and x_2, respectively. m is the number of fired rules (in this example m $= 4$).

Then, the output score for Intel stock is calculated to be 16.63.

Similarly, the output Y value for each stock can be calculated according to the above formula. Then a top 10 list of recommended stocks can be further generated according to the value of Y in descend order.

Step 5: System implementation

The fuzzy web stock information agent, VirtualStreet, is a web-based application suite, which offers functions such as creating new accounts, creating portfolio, editing portfolio, creating top 10 stock list, and so on. The web sever is Jakarta-Tomcat 3.2.1. The database is Microsoft SQL Server 7.0. The applications are written using a variety if languages, including Java, JSP (Java Server Page), HTML (Hyper-Text Makeup Language), Java Servlet, XML and XSL. Fig. 10 shows a screen snapshot the user will see after successfully signing in with correct username and password.

The services provided in the "My Account" section include management tools of portfolio (create, display, edit, and delete), and recommendation list of "Top 10 stocks", which are the precise and reliable results obtained using fuzzy logic method. "Reference Links" provide some useful links on the Internet.

Fig. 11 shows the top 10 stocks determined by the fuzzy logic algorithm implemented in the VirtualStreet, when the user clicks "Top 10 Stocks" on the left menu frame. The information in the list includes ticker symbol, company name, last trade time and price, and trade volume. The most attractive stock is listed first. A timer is set in the application to fetch stock data in order to update the information in the portfolio in a timely fashion. Fig. 12 shows the command interpreter window running in the back end when the web agent fetches stock data from Internet every 5 minutes.

Through this specially designed fuzzy web stock information agent, we can see how a software agent can help us assist the process of decision-making. An intelligent agent employing fuzzy logic technology will further filter and evaluate the degree of importance of the critical information by incorporating uncertainty into reasoning. Such agent can offer the following advantages:

(1). It guarantees a precise classification of the input data by adding a degree of uncertainty, represented by a particular percentage, the fuzzy membership probability;

Fig. 10. The three-frame GUI page displays after user signs in

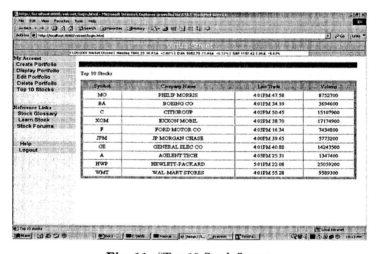

Fig. 11. "Top 10 Stocks" page

(2). Results of the data processing using fuzzy reasoning are precise and accurate, since output values are also computed with the consideration of degree of uncertainty.

(3). Fuzzy Reasoning method is easier to implement than other sophisticated techniques, e.g., neural networks and k-nearest neighbor method, when used for mining large data sets;

(4). The algorithm produces results much faster comparing with other algorithms such as tree based classifiers and ensemble learning;

(5). Fuzzy reasoning algorithm uses fewer number of local variables to obtain the results, which is a big advantage for the systems with low memory;

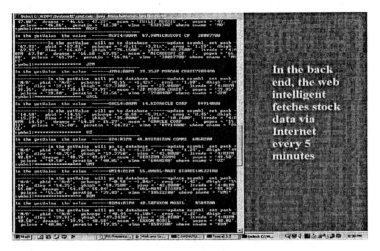

Fig. 12. The web intelligent agent fetches stock data via Internet every 5 minutes

(6). An intelligent agent can help save time when it is instructed to perform certain action on behalf of user.

In the current example, agent can automatically update various information of the stocks in the pre-created portfolio. Since the system is set to fetch the stock data from Internet every several minutes, we can get the latest stock information in a real time manner. Fig. 13 compares the time spent on getting a set of stock information via Internet using the designed web-based agent (VirtualStreet) and using Yahoo! Finance, respectively. It can be clearly seen that the agent shows advantage over Yahoo! Finance when the stock list gets longer.

In today's fast-paced life style, people are faced with tremendous amount of data and information that must be processed and stored. It is already beyond our capability or experience to analyzing all the dynamically-changing data and information and make decision upon those data. We will more and more rely on intelligent automatic ways to perform those decision-making tasks. Thus, agents will be seen employed in more and more applications, given that it can help us in a desired manner, especially in conditions that requires quick action upon dynamically-changing information. In the near future, it is very likely that we can use intelligent homeland security agent that helps identify potential treat or terrorism so that we can prevent disaster, financial agent that helps predict market trend, weather agent that helps predict hazard weather conditions, pest agent helps monitor insects, and lots more.

5 CONCLUSIONS

We are in an era with information overload. It becomes more and more difficult for us to analyze the fast-changing information and make decision upon

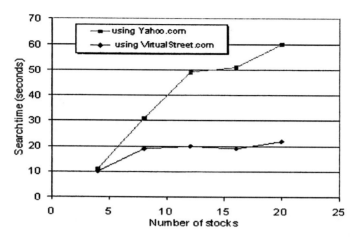

Fig. 13. Searching Time Comparison between VirtualStreet and Yahoo.com [8]

the analyses. Agent - specially designed software - becomes a big help to solve the dilemma. It offers autonomy, and trust and personalization. When equipped with fuzzy logic technology, an intelligent agent can further act on input information or data and assist in making decision or at least providing recommendation. Fuzzy web information classification agent based on FWI is designed as an example, using the latest technologies. The agent can act upon user's instructions and refresh the stock data in a real time manner by accessing the database on the Internet. Using fuzzy reasoning, the agent can create a list of top 10 stocks based on the output values calculated from stock information (as input). With the consideration of degree of uncertainty, results of the data processing are precise and reliable. Through this example, we can the benefits that such intelligent agent can provide us in daily life. In the future, CWI techniques will be used in different smart Web agents.

References

1. Berson A, Thearling K and Smith S (2000) Building data mining applications for CRM. McGraw Hill Osborne Media
2. DuBois D and Prade H (1997) Fuzzy sets and systems: theory and applications. Academic Press, New York
3. Elmasri RA and Navathe SB (2000) Fundamentals of database systems. Addison-Wesley, New York
4. Kandel A (1986) Fuzzy mathematical techniques with applications. Addison-Wesley Publishing
5. Agnes M (1999) Webster's New World College Dictionary 4th edn. Macmillan, New York
6. Smithsin M (1989) Ignorance and uncertainty: emerging paradigms. Springer-Verlag, New York

7. Wang L-X (1997) A course in fuzzy systems and control. Prentice Hall, Upper Saddle River, New Jersey
8. Wang Y (2001) Personalized real time stock information agent using fuzzy data mining method. M.S. thesis, Georgia State University
9. Wang Y, Zhang Y-Q, Belkasim S and Sunderraman R (2002) Real time fuzzy personalized Web stock information agent. In:Proc. of ISDA 2002, pp 83-88
10. Witten IH and Frank E (2000) Data mining, practical machine learning tools and techniques with Java implementations. Morgan Kaufmann Publishers
11. Zadeh LA (1965) Fuzzy sets. Information and Control 8:338–353
12. Zadeh LA (1986) Fuzzy algorithms. Information and Control 12:94–102
13. Zhang Y-Q, Akkaladevi S, Vachtsevanos G and Lin TY (2002) Granular Neural Web Agents for Stock Prediction. Soft Computing Journal, 6:406–413
14. Zhang Y-Q, Fraser MD, Gagliano RA and Kandel A (2000) Granular neural networks for numerical-linguistic data fusion and knowledge discovery. In: Special issue on neural networks for data mining and knowledge discovery, IEEE Transactions on Neural Networks 11:658–667
15. Zhang Y-Q and Lin TY (2002) Computational web intelligence (CWI): synergy of computational intelligence and web technology. In: Proc. of FUZZ-IEEE2002 of world congress on computational intelligence 2002: special session on somputational web intelligence, 1104-1107